Modelling and Optimisation of Fluid–Particle Processes in Environmental Engineering

This comprehensive book explores fluid–particle interaction processes in environmental engineering, combining theoretical foundations with advanced numerical simulation techniques. Covering eight key areas from particle agglomeration to membrane fouling, the work integrates fundamental physics with practical computational tools to address critical challenges in environmental systems and reactor design.

It provides a detailed overview of hydrodynamic modelling, interparticle forces, and membrane separation dynamics, alongside practical numerical simulation approaches for real-world applications. It also delivers essential insights into solid–liquid interface processes, mathematical modelling for fouling prediction, and design optimisation strategies for stirred tank reactors, rotatory vortex pelleting systems, and hydrothermal carbonisation reactors. Readers will gain access to proven methodologies that bridge theoretical understanding with practical implementation, offering workflow simplification tools for complex environmental engineering challenges.

Ideal for both academic study and professional engineering practice, this essential resource targets graduate students and researchers in environmental and chemical engineering, along with practicing engineers in water treatment, membrane technology, and reactor design.

Modelling and Optimisation of Fluid–Particle Processes in Environmental Engineering

Edited by
Benjamin Oyegbile

CRC Press
Taylor & Francis Group
Boca Raton London New York

CRC Press is an imprint of the
Taylor & Francis Group, an **informa** business

Designed cover image: unsplash.com

First edition published 2026
by CRC Press
4 Park Square, Milton Park, Abingdon, Oxon, OX14 4RN

and by CRC Press
2385 NW Executive Center Drive, Suite 320, Boca Raton FL 33431

British Library Cataloguing-in-Publication Data
A catalogue record for this book is available from the British Library

ISBN: 978-1-032-85489-2 (hbk)
ISBN: 978-1-032-85490-8 (pbk)
ISBN: 978-1-003-51842-6 (ebk)

DOI: 10.1201/9781003518426

Typeset in Times
by Deanta Global Publishing Services, Chennai, India

Access the Support Material: www.routledge.com/9781032854892

This book is dedicated to the loving memory of our parents

Abraham Oladepo Oyegbile

Peres Nekesa Wandera

Contents

Preface

Fluid–particle systems form an integral part of many environmental and process engineering applications. These systems are widely used in many engineered environmental processes ranging from water and wastewater treatment systems, solid–liquid separation reactors, to urban drainage systems. The fluid–particle processes involve an interplay of fluid flow forces and physicochemical interactions such as fluid–particle, particle–particle, and particle–wall interactions. This makes the fluid–particle processes complex in nature and difficult to analyse, predict, and optimise.

For complex processes such as fluid–particle interactions, studies of real-world systems are being carried out using modelling and simulation tools for easier and accurate representation of the different phenomena involved in such complex systems. Mathematical modelling and numerical simulations play a key role in predicting, analysing, and minimising operational problems associated with these complex processes. In addition, these mathematical and numerical models are essential in optimising the performance of processes and systems. In light of the significance of studying real-world systems from the first principles, we present a book on *Modelling and Optimisation of Fluid–Particle Processes in Environmental Engineering*, which will be an essential resource for students, researchers, and engineers who are dealing with complex fluid–particle systems, particularly in the environmental and process engineering field. Through theoretical analysis and case studies, the book aims at giving a detailed understanding of how models for fluid–particle systems can be developed and used for various purposes such as prior prediction of fluid–particle processes, simulation, and performance optimisation as well as minimisation of operational problems.

First, we present some theoretical discussions on the fundamentals of fluid dynamics, fluid–particle interactions, and their effects in fluid–particle systems. The first half of this book deals with the basic phenomena that occur in fluid–particle systems. Chapter 1 introduces the reader to fluid dynamic forces in the agglomeration of fine particles. Chapter 2 discusses the effects of fluid dynamics and fluid–particle interaction on membrane-based separation in wastewater treatment. This chapter is intended to show the key roles played by various forces in fluid–particle systems. Chapter 3 provides a theoretical discussion on various microprocesses occurring at solid–liquid interfaces in fluid–particle systems. And lastly, Chapter 4 discusses the fundamentals of numerical simulation and physicochemical characterisation of fluid–particle systems.

The second half of the book presents case studies on modelling, simulation, and optimisation of various fluid–particle systems. Chapter 5 deals with a case study of numerical simulation and optimisation of hydrodynamics and mixing processes in stirred tank reactor. Chapter 6 presents a case study on the modelling of hydrodynamics and fluid–particle interactions in a rotary vortex pelleting reactor. Another example on the modelling of convective mixing and heat transfer in a hydrothermal carbonisation reactor is presented in Chapter 7. Finally, Chapter 8 presents a modelling case study involving a recently developed mathematical model for membrane-based separation systems. These case studies provide a starting point for students and researchers who want to model and optimise fluid–particle systems. They will also enhance the learning process and provide examples for beginners to comprehend the basic steps involved in their modelling of fluid–particle systems. We sincerely hope that this contribution will be found useful as a teaching material for students and a reference book for practicing engineers looking to simplify their workflow.

Acknowledgements

This textbook which is based on many years of rigorous research work would not have been possible without the support, collaboration, encouragement, and expertise of many individuals and organisations.

First and foremost, we would like to acknowledge the immense contributions of our colleagues and mentors to the realisation of this book project. We would especially like to thank Professor Guven Akdogan and Professor Lingam Pillay, especially for their academic mentorship and pastoral care during the period of our intense research work at Stellenbosch University. Their intellectual contributions were pivotal in shaping the structure and depth of this book.

We would also like to acknowledge the support, enthusiasm, and constructive critique of some of our colleagues at the Department of Process Engineering at Stellenbosch University where most of the research work in this book was conducted. In particular, we would like to thank Dr Mohsen Karimi and Mr Nelson Kyakutwika for their vital contribution to our studies on numerical simulations and mathematical modelling of fluid–particle interactions in vortex and membrane reactors.

We are also deeply grateful to our family and friends for their support, encouragement, and understanding throughout the period of writing this book. Their constant belief in us provided the strength and sense of purpose needed to bring this work to fruition.

Last but not least, we would like to express our gratitude to the World Academy of Sciences (TWAS) and the National Research Foundation (NRF) for their generous funding to carry out some of the research work reported in this book.

Benjamin Oyegbile
Suzanne Yala

Editor

Benjamin Oyegbile holds a PhD in Environmental Engineering and is currently a visiting researcher at the University of Bradford. He has over six years of university-level teaching experience and a focused research agenda in fluid–particle modelling and environmental fluid dynamics. His research work is focused on the integration of experimental analysis and high-fidelity numerical simulations with advanced machine learning techniques to investigate and optimise complex fluid–particle processes in environmental engineering.

Contributor

Suzanne Yala is a lecturer and a chemical engineer at the Department of Chemical Engineering in the Dedan Kimathi University of Technology. She graduated from Stellenbosch University with a PhD in Chemical Engineering. Her research interests are focused on water and wastewater treatment, decentralised sanitation systems, and membrane technology applications and modelling.

The Role of Hydrodynamics and Interparticle Forces in the Agglomeration of Fine Particles

1

Benjamin Oyegbile

1.1 INTRODUCTION TO COLLOIDAL STABILITY AND INTERFACIAL FORCES

Fluid–particle interactions play an important role in many natural and engineered processes. Hence, it is important to understand the fundamentals of the fluid–particle interactions and the forces that govern these interactions. Particles are ubiquitous in nature and drive many atmospheric and hydrodynamic processes, such as air pollution and sediment transport. Particles also play vital roles in many environmental process industries such as water and wastewater purification and urban drainage systems. A number of interfacial forces and interactions that play key roles in particle interactions are well understood in the light of classical and extended Derjaguin–Landau–Verwey–Overbeek (DLVO) theories [1–5], and therefore, it provides a basis for any theoretical analysis of particle dispersion and colloidal stability.

Most colloidal particles in suspension (0.001–10 µm) tend to remain suspended and settle very slowly due to the presence of surface charge arising from isomorphic substitution, chemical reaction at the interface, or preferential adsorption of ions on the particle surface from the surrounding medium [6, 7]. The charge on the particle surface is surrounded by an excess of oppositely charged ions (counterions) in solution in order to maintain electrical neutrality. The combined interactions of this system of opposite charge clouds are known as the electric double layer [8]. The developed electrostatic repulsive forces prevent particle aggregation in colloidal systems and hence contribute to their stability in the dispersed system [9]. The diffuse layer and the Stern layer of the charge cloud surrounding a particle together constitute the electric double layer [10].

A quantitative description of the colloidal stability and aggregation process was facilitated by the assumptions of the DLVO theory, which considered interactions between particles as additive [8]. The

DOI: 10.1201/9781003518426-1

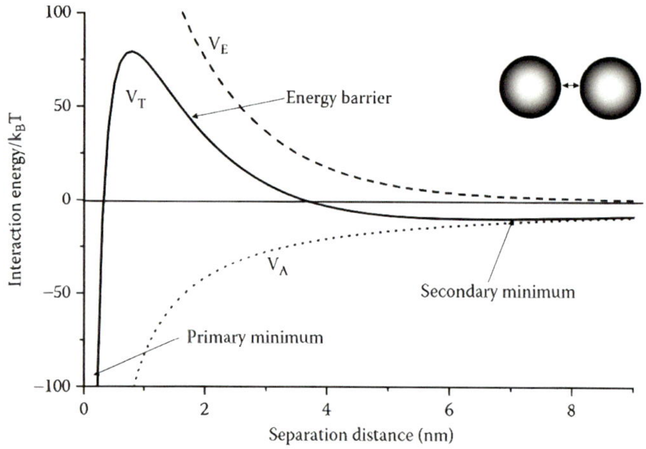

FIGURE 1.1 Potential energy barrier in the interaction between two particles. Reprinted with permission from CRC Press [17].

DLVO theory was primarily based on two types of interactions between two compact spherical particles—electrostatic (including *Born repulsion*) [11] repulsive force due to the electrical double layer and the Van der Waals attractive force of the dispersion. The combination of these two interactions defines a total energy barrier V_T or the so-called *Gibbs interaction energy,* as shown in the energy profile in Figure 1.1 [5, 8, 12–15] which, if overcome by the kinetic energy of the moving particles, will result in aggregation. This is expressed mathematically in Equations (1.1) and (1.2), where V_A, V_R, and ζ refer to the energy barrier due to Van der Waals' attraction and electrostatic repulsion and zeta potential, respectively. In the case of submicron particles, the kinetic energy is derived primarily from Brownian diffusion and less from hydrodynamic (fluid shear) or gravitational forces (differential settling). However, for larger or coarser particles, hydrodynamic interactions such as turbulence or fluid shear and gravitational forces impart much of the kinetic energy needed to overcome the flocculation barrier [16, 13, 17]. A critical shear rate is required to overcome the energy barrier in order to allow flocculation and floc formation, and this is dependent on the surface charge and the size of particles. In general, the higher the charge on the particles, and the smaller the particle size, the higher the shear rates that are required to bring about flocculation [1, 15, 18]:

$$V_T = V_R + V_A \tag{1.1}$$

$$V_T = \pi\varepsilon\zeta^2 d\exp(-kh) - \frac{Ad}{24h} \tag{1.2}$$

In addition to the interactions described earlier, there exist additional interaction energies and structural forces which may arise, for instance, from the perturbation or re-arrangement of water molecules near the interface [14, 19, 20]. Non-DLVO interactions such as hydrophobic, hydration, oscillatory, structural, steric, and electrosteric and hydrodynamic forces, which were previously not considered in colloidal interactions, are now widely recognised to play important roles in colloidal aggregation [1, 20–23]. These forces may be attractive, repulsive, or oscillatory in nature and may be more pronounced or stronger than the DLVO forces [20]. In some cases where the energy barrier for aggregation exists at sub-nanometre distances, significant discrepancies between theoretical (DLVO-based) and experimental measurements of flocculation rates have been reported [16, 24]. For instance, in the case of colloidal dispersion containing clay particles, the theory provides a general conceptual model for the gross interactions of particles but fails to explain certain aspects of the interactions [7]. However, in most practical applications of flocculation, the dominant forces are electrostatic repulsion and London–Van der Waals attraction forces [25].

1.2 FLOCCULATION AND KINETICS OF PARTICLE AGGREGATION

In the environmental process industry, the removal of particulate solids from a dispersed phase is of great importance. However, when the sizes of solid particles diminish and reach micron and submicron ranges, the particles tend to remain in suspension and cannot be removed by gravity settling alone. In order to achieve an acceptable solid–liquid separation at a reasonable cost, the particles need to be agglomerated by flocculation subsequently followed by phase separation processes such as sedimentation, decantation, filtration, and centrifugation [26].

Flocculation plays a key role in several natural processes in the aquatic and marine environment, such as sediment erosion, transport, aggregation, and deposition, and it is often employed in several industrial purification and separation processes [20, 27–32]. Aggregation of cohesive suspended particles by size enlargement to form larger flocs is typically encountered in natural systems, such as rivers, reservoirs, lakes, and estuaries, and in engineered systems such as bioreactors and agitation vessels [24, 33–36]. It is widely employed in both upstream and downstream solid–liquid separation processes where the value of the individual phases is enhanced by destabilisation and aggregation of the charged particles in suspension using high molecular weight synthetic polymers [37–41]. Typical mean shear rates in these diverse natural and engineered systems range from 2.5 s^{-1} in natural systems to 5000 s^{-1}, especially in high shear bioreactors and mixing chambers [42, 43].

Flocculation kinetics deals with time-dependent changes in the dispersions or suspensions and provides information on the flocculation rate, dispersion stability, and particle interactions, which depend on the number and efficiency of the collisions [44]. The collisions between particles have been suggested to occur at a rate that depends on the transport mechanisms of Brownian diffusion, fluid shear, and differential settling, and it is assumed to be a second-order rate or reaction process [45]. The aggregation of dilute systems (TS $\leq$ 1%) with an initial concentration of primary particles n_1 and fast aggregation constant k^f can be represented by the expression in Equation (1.3) [11, 45]:

$$\frac{dn_1}{dt} = -k^f n_1^2 \tag{1.3}$$

Furthermore, several studies have shown that aggregates formed from different transport mechanisms exhibit marked variations in their structural attributes. The aggregates formed by shear-induced collision are known to be stronger than those formed from other transport mechanisms, those from Brownian motion tend to be easily dispersed by shearing, and those from differential settling result in ragged, weak, and low-density aggregates [46]. Smoluchowski, using the 'population balance' approach, expressed time evolution of the number density of discrete particles of size k as they aggregate with respect to time in terms of the collision frequency or rate function β originally for Brownian diffusion and laminar or uniform shear flow and later for differential settling, assuming binary collisions between the particles [21, 45, 47–49]. Although the Smoluchowski equation has been modified in line with recent findings to include several other parameters, the general form of the expression is presented in Equation (1.4). The quantity β_{ij} is the collision frequency or rate function for collisions between ith- and jth-sized particles while α_{ij} is the dimensionless collision efficiency factor. The collision frequency depends on the transport mechanisms of Brownian motion, fluid shear, and differential settling, whereas the collision efficiency is a function of the degree of particle destabilisation and it gives the probability of collision leading to attachment with values ranging from 0 to 1 [47, 50–53]:

$$\frac{dn_k}{dt} = \frac{1}{2}\sum_{i+j=k} \beta(i,j)n_i n_j - \sum_{i=1}^{\infty} \beta(i,k)n_i n_k \tag{1.4}$$

The first term of the right-hand side of Equation (1.4) describes the increase in the number of particles of size k by flocculation of two particles whose total volume is equal to the volume of a particle of size k, while the second term on the right-hand side describes the loss of particles of size k due to aggregation with particles of other sizes. The general form of the equation expresses the rate of change in the number concentration of particles of size k. In arriving at Equation (1.4), Smoluchowski made a number of key assumptions listed below:

- The collision efficiency factor α is unity for all collisions.
- The fluid motion undergoes laminar shear.
- The particles are monodispersed.
- There is no breakage of flocs.
- All particles are spherical in shape before and after collision.
- Collision involves only two particles.

1.3 FLOCCULATION TRANSPORT PROCESSES

The interparticle collisions are promoted by transport mechanisms of Brownian motion or perikinetic (for particles with diameters less than 1 µm), fluid shear or orthokinetic (for particles in the diameter range 1–40 µm), and differential settling (typically for particles with diameter larger than 40 µm). The collision frequency or rate function β for these three mechanisms can be expressed mathematically in Equations (1.5)–(1.7), where $\beta_{BM,}$ $\beta_{SH,}$ and β_{DS} refer to the collision frequency function for Brownian motion, fluid shear, and differential settling, respectively. The total collision rate or frequency β_T is the sum of contributions from each of the three transport mechanisms (Equation 1.8) [28, 29, 34, 46, 48, 51, 54–56]:

$$\beta_{BM} = \frac{2kTd_{ij}^{2}}{3\mu d_i d_j} \tag{1.5}$$

$$\beta_{SH} = \frac{(d_i + d_j)^3}{6} G \tag{1.6}$$

$$\beta_{DS} = \frac{\pi g}{72\nu\rangle_w}(d_i + d_j)^2(\Delta\rho_i d_i^2 - \Delta\rho_j d_j^2) \tag{1.7}$$

$$\beta_T = \beta_{BM} + \beta_{SH} + \beta_{DS} \tag{1.8}$$

1.3.1 Perikinetic Particle Aggregation

The random displacement of particles in Brownian motion as a result of the thermal energy of the system is termed perikinetic [1, 21, 23, 45]. In practical applications, perikinetic aggregation is only important for very small particles where they are continuously bombarded by the surrounding water molecules (<1 µm) [57]. Under this prevailing condition, the rate of floc growth cannot be sufficiently sustained for an effective phase separation. Hence, the importance of other transport mechanisms in promoting floc growth kinetics. However, in the case of contact between a small p article and microflocs, Brownian diffusion can still control the transport of small particles across the layer of fluid on the floc surface [58].

1.3.2 Classical Orthokinetic Aggregation

The agitation of suspended particles in a liquid medium will lead to aggregation due to the interparticle collision induced by the fluid motion [17]. The velocity gradient or shear rate is dependent on the nature of the fluid flow. In uniform laminar shear flow, the velocity gradient remains constant in the entire flow field, while in turbulent flow, there is a rapid fluctuation of the velocity gradient. Consequently, the velocity gradient is a function of both space and time (*spatial and temporal*) [59, 60]. Camp and Stein, in their critique of the Smoluchowski approach to shear flocculation modelling [61], introduced the concept of root mean square (RMS) or absolute velocity gradient $\bar{G}$ to account for the variations in the shear rate [6, 46, 58, 62, 63]. Considering the angular distortion of an elemental volume of fluid arising from the application of tangential surface forces, $\bar{G}$ is defined according to Equations (1.9) and (1.10) as the RMS velocity gradient in the mixing vessel, where u, v, and w are the components of the fluctuating velocity while x, y, and z refer to the 3D Cartesian coordinate system, P is the power dissipated, V is the volume of the mixing vessel, ε is the energy dissipation rate per unit mass, μ is the dynamic viscosity, and ν is the kinematic viscosity [6, 64]:

$$G=\left\{\left(\frac{\partial u}{\partial y}+\frac{\partial v}{\partial x}\right)^2\left(\frac{\partial u}{\partial z}+\frac{\partial w}{\partial z}\right)^2\left(\frac{\partial v}{\partial z}+\frac{\partial w}{\partial y}\right)^2\right\}^{\frac{1}{2}} \tag{1.9}$$

$$G=\frac{1}{\tau_{Kolmogorov}}=\sqrt{\frac{P}{\mu V}}=\sqrt{\frac{\varepsilon}{\nu}} \tag{1.10}$$

In practical applications, owing to the difficulties associated with the calculation of absolute velocity gradient $\bar{G}$ due to the fluctuations in the energy dissipation within the mixing vessel, an average velocity gradient G_{ave} has often been used in place of the absolute value $\bar{G}$ [59, 60] (Equation 1.11). Korpijarvi et al. [65], in their study of mixing and flocculation in a jar using computational fluid dynamics (CFD), reported a large variation in the local velocity gradient G_L within the mixing vessel. In similar studies conducted by Kramer and Clark [6, 66] as well as by Mühle [67], a lower estimate of the absolute velocity gradient was proposed (Equations 1.12 and 1.13) to account for the variance of the velocity gradient throughout the fluid where quantities G_L, P_{ave}, ε, V, μ, and ν represent the local velocity gradient at different points within the mixing vessel, average power consumption, volume of the mixing vessel, kinetic energy dissipation rate, dynamic viscosity, and kinematic viscosity, respectively:

$$G_{ave}=\sqrt{\frac{P_{ave}}{\mu V}}-\sqrt{\frac{\varepsilon_{ave}}{\nu}} \tag{1.11}$$

$$\bar{G}=\sqrt{\frac{P}{\mu V}-\sigma_{GL}^2} \tag{1.12}$$

$$G=0.26\sqrt{\frac{P}{V\mu}}=0.26\sqrt{\frac{\varepsilon}{\nu}} \tag{1.13}$$

1.3.2.1 Particles in Laminar Shear

According to the rectilinear model of particle trajectory, two particles moving in a fluid on different streamlines will experience a velocity gradient, which indicates the relative motion of particles and the possibility of collision and aggregation [45, 68]. In simple laminar or uniform shear with a well-defined flow field such as in pipe flow or in some Couette devices [42, 69, 70], the transport of particles by laminar shear due to the fluid motion can be characterised by a single value of the shear rate G [6]. Under such conditions, the particles in laminar shear will exhibit rotational motion in the direction of travel with a constant angular velocity ω [71, 72].

1.3.2.2 Particles in Turbulent Shear

In most practical applications of flocculation, the aggregation processes typically occur under turbulent conditions [6, 68, 73–75]. The particles in suspension under the influence of turbulent flow will experience fluctuating motion of the fluid with the particles being transported by the fluid eddies and the flow vortex [6, 58]. Thus, small particles suspended in fluid exist in an environment of small energy-dissipating eddies in most typical sheared reactors [42, 76]. Under such conditions, particle collisions are promoted by eddy size similar to that of the colliding particles [47, 77]. The mechanism of flocculation in turbulent shear has been suggested to be similar to that of laminar shear for particles smaller than the Kolmogorov microscale λ_0 of turbulence, while for particles larger than this length scale, the flocculation mechanism is similar to that of Brownian diffusion [1, 6].

The formation of aggregates in any turbulence or shear-induced flocculation essentially consists of the following phases: destabilisation; collision and adhesion; floc growth and deformation phases [78]. Ives [69] and Bergenstahl [79] expressed the kinetics of orthokinetic aggregation in terms of the change in the initial particle concentration, while Levich [62, 80] in an earlier study derived the collision rate equation based on the concept of locally isotropic turbulence assuming a viscous diffusive subrange (i.e., particle size smaller than Kolmogorov microscale) for shear-induced turbulent flow where N, r, and d represent the particle number concentration per unit volume, particle radius, and diameter, respectively (Equations 1.14 and 1.15). Several other models of flocculation kinetics are available in the literature [7, 81]. In most of these models, the shear rate G and the kinetic energy dissipation rate ε are shown as important drivers of the flocculation process. Typically, orthokinetic flocculation can be experimentally observed in a conventional stirred tank consisting of an axially mounted impeller and a circular or rectangular mixing vessel:

$$\frac{-dN}{dt_{(\text{Couette})}} = \frac{2}{3} f_o G d^3 N^2 + \frac{4}{3} f_p \frac{kTN^2}{\mu} \tag{1.14}$$

$$\frac{dN}{dt_{Turb.}} = -6\pi\beta\sqrt{\frac{\varepsilon_{\text{intensity}}}{\eta}}\, r^3 N^2 = -12\pi\beta\sqrt{\frac{\varepsilon_{\text{intensity}}}{\eta}}\, d^3 N^2 \tag{1.15}$$

1.3.2.3 Extended Orthokinetic Aggregation

Pelleting flocculation is considered an extension of the classical orthokinetic transport mechanism based on the ‘metastable state’ concept [82, 83]. It has been shown by several studies that the efficiency of the floc formation process as well as the floc structural attributes (size, shape, porosity, density, etc.) can be significantly improved by the application of suitable mechanical energy [84–94]. This is accomplished by the application of a non-destructive uneven force system on the floc surface, by the action of the turbulent fluid motion, or the so-called floccule mechanical synaeresis induced by rolling and collision mechanisms [95–99].

In the rolling mechanism, as the floc rolls along a plane surface, it is subjected to a pressure fluctuation which becomes greater at the front of the central point in the contact area and lesser at the rear. On the other hand, in the collision mechanisms, a floc experiences an impact due to the floc–floc or floc–surface collision. In order to withstand this impact, the floc must adhere strongly, in other words, the impact results in a more compact floc [83]. The resulting pressure induced in a curvilinear trajectory on the floc surface can be expressed as a function of the velocity gradient G where m, R, G, and A represent the mass of the floc, the radius of the flow trajectory, velocity gradient, and the area of the floc projection onto a

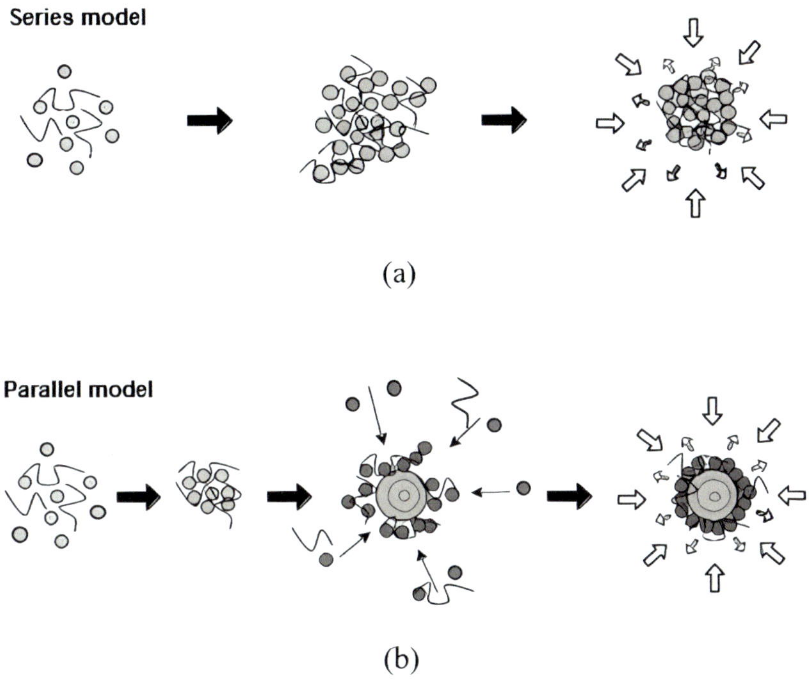

FIGURE 1.2 Conceptual models of pelleting flocculation process: (a) series model—raspberry structure and (b) parallel model—onion structure.

plane oriented normally to the vector of the resultant force, respectively (Equation 1.16). In practice, pelleting flocculation is realised through the selection of an appropriate process engineering condition and stirrer–vessel system (geometry and configuration) in order to obtain the necessary conditions suitable for rolling and collision-mediated floc palletisation [83, 91]:

$$P = \frac{mRG^2}{A} = \frac{mRv^2}{A\lambda_0^4} \tag{1.16}$$

The structure formation by pelleting flocculation yields 'pellet flocs' with superior structural attributes when compared to the 'random flocs' produced by the classical flocculation process, which tend to be loose and bulky [84]. Previous studies have reported two distinct models of pelleting flocculation, namely parallel and series models resulting in onion and raspberry-like pellet floc structures as presented schematically in Figure 1.2 [87, 90, 91, 100–103]. In the series model, orthokinetic flocculation occurs first, where random, loose, and bulky flocs are produced. These voluminous flocs shrink and densify by the application of suitable mechanical energy (mechanical synaeresis).

Conversely, in the parallel model, there is no time delay between the orthokinetic flocculation and mechanical synaeresis. The dispersed particles in suspension and micro flocs are transported by the fluid motion to the surface of the 'mother seed' where they are attached by polymer bridging [95, 100, 102]. In terms of the flocculation efficiency and process-relevant (phase separation) parameters, pellet flocs can be characterised by their size, shape, density, porosity, compressive strength, and deformability.

1.3.3 Polymer-Mediated Interactions

The energy barrier to flocculation can be overcome in several other ways apart from the kinetic energy derived from the transport mechanisms. The electrostatic charge can be lowered with the aid of polymeric

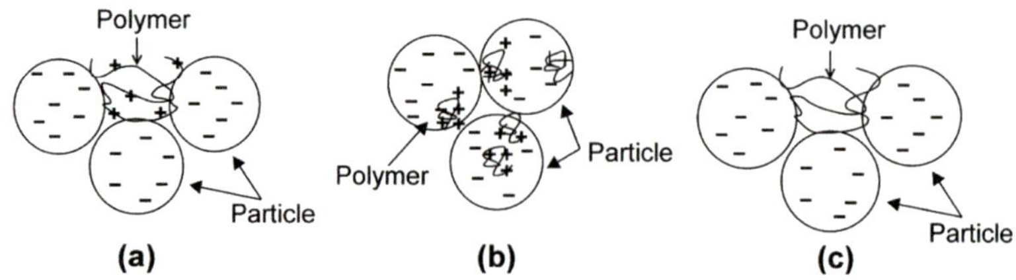

FIGURE 1.3 A schematic representation of the dominant mechanisms in polymer-mediated interactions: (a) charge neutralisation, (b) charge patch formation, and (c) polymer bridging.

flocculants by double-layer compression, charge neutralisation, and polymer bridging between particles [12, 13, 104]. In practice, both of these approaches are employed either concurrently or sequentially. In addition to these interactions, several other mechanisms have also been attributed to the polymer flocculation of charged particles in suspension. Electrostatic charge patch; depletion flocculation; network flocculation, and polymer complex formation are a few of other relevant interactions [16, 18, 20, 27, 73, 105–108].

However, the dominant concepts in the case of polyelectrolyte-mediated destabilisation are surface charge neutralisation (ion exchange), charge patch formation, and polymer bridging as illustrated in Figure 1.3 [44, 109–115]. The important physicochemical properties that influence the flocculation process in the case of polymer-mediated interactions include polymer molecular weight, charge density and concentration, zeta potential or surface charge of suspension, particle size and distribution, specific surface area, solution conductivity, and pH [105]. In the case of pelleting flocculation, Higashitani and Kubota [98] in their study of the pelleting flocculation of polystyrene latex particles reported that the formation of pellet flocs depends on the suspension concentration, flocculant molecular weight, and charge density as well as the intensity of mixing. They argued that pellet flocs are formed within a limited range of these parameters, hence pelleting flocculation requires a high degree of control of different process conditions. On the basis of their findings, they expressed mathematically the volume fraction of particles ϕ_{pt} at the critical particle concentration N_{c} above which pellet flocs are formed as a function of the initial particle concentration N_0, polymer concentration C_{p}, charge density p, and the molecular weight M, where N_{A} is the Avogadro constant (Equation 1.17):

$$\phi_{pt} = 4\times10^{-5}\frac{C_p N_A}{pN_0 M} = 24.08\times10^{18}\frac{C_p}{pN_0 M} \tag{1.17}$$

1.4 HYDRODYNAMICS OF FINE PARTICLE AGGREGATION

Hydrodynamically induced turbulent shear is an important driver of the flocculation process, especially in the case of *orthokinetic* aggregation of particles [13, 16, 116]. The floc growth and stability in any flocculated system has been suggested to be a function of fluid–particle interactions and the intrinsic physicochemical properties of the floc [75, 77, 117]. The dynamics of these interactions affects all facets of the flocculation process and the degree of aggregation in any sheared system [118]. In the case of hydrodynamic interactions, induced velocity gradient promotes the aggregation process but might also be responsible for floc breakage as a result of increased viscous shear stress [119]. Consequently, in the case of shear-induced collisions, the hydrodynamic effect can be very significant [57].

Chemical transport in natural and engineered systems is often controlled more by the fluid properties, in this case the mixing rate of the fluid, than by chemical properties such as the molecular diffusivity of the compound. Therefore, in the design of flocculation reactors, it is desirable to have uniform mixing

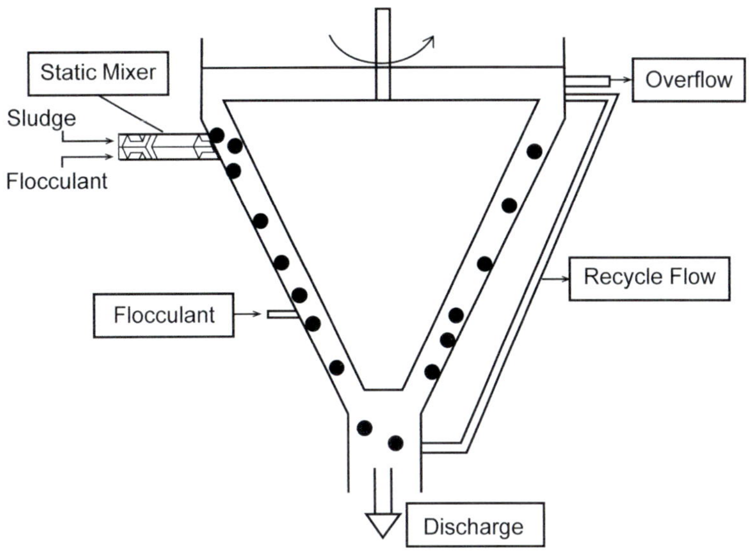

FIGURE 1.4 Schematic diagram of a pelleting reactor showing the role of mixing in the structure formation process.

intensities throughout the reactor [6]. Under the condition of shear-induced turbulence due to the use of flow inducers such as paddles, mixers, stirrers, or bubbles (Figure 1.4), the convective currents in the system are insignificant as compared to the advective currents generated in the system [42]. The mathematical modelling of the phenomena of fluid–particle mixing, floc formation, growth, and breakup is very complex; as such, minimising the energy dissipation rate (degree of turbulent mixing) in the reactor will lead to improvement in flocculation efficiency [6].

1.5 FLUID MIXING AND PARTICLE DISPERSION

Turbulent mixing characterised by high Reynolds number ($R_e > 10^4$) [120] can be viewed as a hierarchy of irregular, rotational, and dissipative motion containing vorticity (curl or rotation of the velocity vector) on different scales or eddy sizes [5, 121]. In turbulent flows such as those encountered in pipe or tube flow, channel flow, and stirred tanks or mixing chambers, energy transfer occurs on different eddy scales [64, 67]. Eddies are spatially recognisable flow patterns that exist in a turbulent flow for at least a short time in which there is a correlation between the velocities at two different points [6]. Fluid deformation causes vortices to stretch and vorticity and kinetic energy to be transported from larger to smaller eddies [5, 121]. The turbulent vortex in such fluid motion is generally propagated in the tangential and axial direction rather than in the radial direction with the vorticity increasing with decreasing eddy size [64, 121, 122].

In most practical applications of mixing, there exist three scales of mixing, namely macromixing, mesomixing, and micromixing. Macromixing refers to mixing that is driven by the largest scale of motion in the fluid Λ (integral length scale), mesomixing on the other hand involves mixing on a scale smaller than the bulk circulation but larger than the micromixing, while micromixing is the mixing on the smallest scale of fluid motion λ_0 (Kolmogorov microscale) expressed in Equation (1.18) and at the final scales of molecular diffusivity (Batchelor scale) [120, 123–126]. Micromixing describes the process of homogenisation of liquid balls with their surroundings on a molecular level. In typical mixing conditions, the dividing line between micro- and macroscale is between 100 and 1000 μm, respectively [127]:

$$\lambda_0 = \left[\frac{\nu^3}{\varepsilon}\right]^{\frac{1}{4}} \tag{1.18}$$

The largest eddies in turbulent dispersion that represent the macroscale of turbulence or the integral length scale (Λ) contain most of the energy and are produced by the stirrer or agitator head. The size of this macroscale of turbulence is of the order of the diameter of the agitator [42, 53, 120]. The turbulent flow can be viewed as an eddy continuum, with its size ranging from the dimension of the turbulence generating device to the Kolmogorov length scale [53]. In between the energy-containing eddies of the integral length scale Λ at the upper end of the inertia subrange and the smallest eddies of Kolmogorov microscale λ_0 at the beginning of the viscous dissipation range [128], there exist many eddies of other scales smaller than the integral scale Λ that transfer kinetic energy continually through the other length scales. The Batchelor λ_B and Taylor λ_T scales expressed in Equations (1.19) and (1.20) are examples of other important length scales, where D is the diameter of the flocs while ε is the turbulent kinetic energy dissipation rate:

$$\lambda_B = \left[\frac{D^2\nu}{\varepsilon}\right]^{\frac{1}{4}} \tag{1.19}$$

$$\lambda_T = \frac{u\sqrt{15}}{\sqrt{\frac{\varepsilon}{\nu}}} \tag{1.20}$$

The Taylor scale λ_T is an intermediate length scale in the viscous subrange that is representative of the energy transfer from large to small scales, but it is not a dissipation scale and does not represent any distinct group of eddies [129]. Batchelor microscale, on the other hand, is a limiting length scale where the rate of molecular diffusion is equal to the rate of dissipation of turbulent kinetic energy and represents the size of the region within which a molecule moves due to diffusional forces [123, 125, 128, 130].

1.6 ENERGY DISSIPATION IN TURBULENT FLOW

The power input into any agitated system will eventually dissipate as heat due to viscous forces [76]. The larger eddies of the order of the agitator diameter ($\Lambda \sim d_A$) draw energy from the fluid motion, while the smaller eddies transfer that energy gradually and continually to the smallest eddy, where the energy is ultimately dissipated into heat by friction [7]. The turbulence parameters of interest in any agitated microsystem (with respect to the mixer capacity and performance) are the Kolmogorov microscale λ_0, turbulence kinetic energy dissipation rate ε, and agitator tip velocity V_{tip} [122, 124, 131].

The Kolmogorov length scale λ_0 is an indicator of the rate of micromixing and mode of agglomerate deformation due to turbulent shear, while the energy dissipation rate ε is the rate of dissipation of kinetic energy. The tip velocity V_{tip} is a measure of the tangential velocity imparted by the flow inducer, an indicator of the strength of the vortex generated by the flow. Peripheral velocity V_{pkol}, on the other hand, gives an indication of the velocity of the flow vortex at Kolmogorov microscale [76, 122, 124, 131]. In typical laboratory mixing experiments ($D < 1$ m), the microscale of turbulence predominates. Under such conditions, particle collision is promoted by eddy size similar to that of the colliding particles [47].

1.7 AGGREGATE STABILITY IN SHEARED SYSTEM

Floc stability under the influence of hydrodynamic force has been suggested to be a function of floc binding or cohesive force F_B and hydrodynamic breaking force F_H [64, 81, 116, 132–135]. While the binding force is determined by the flocs' structure and physicochemical attributes, flow turbulence is the principal factor in the case of hydrodynamic force [59, 75, 117]. Of these two governing factors of floc stability, turbulence is the least understood owing to its complex nature [45, 136]. Therefore, a detailed analysis of floc stability under turbulent conditions often encountered in natural and agitated systems is not only difficult but often time consuming [45].

Recent advances in computational fluid dynamics have vastly improved the understanding of turbulence phenomena. However, numerical modelling of complex turbulent fluid–particle interactions remains a challenge in fluid mechanics, and it is computationally intensive [122]. In addition, the assumption of the rectilinear motion of particles prior to collision in flocculation modelling has been supplanted by recent findings of curvilinear particle trajectories prior to collisions [47, 63]. Particles have been shown to move around other particles in a curvilinear path, thereby reducing the probability of collision and attachment [47, 57].

1.7.1 Aggregate Strength and Hydrodynamic Stress

Floc formation and growth process in turbulent flocculation comprise the lag phase, swift growth phase, and steady-state phase or breakup/restructuring phase [135, 137, 138]. The overall rate of floc growth at the initial formation phase up to the steady-state phase has been suggested to be a balance between collision-induced particle aggregation and floc cohesive strength, on the one hand, and the rate of floc breakage due to hydrodynamic stress, on the other hand [64, 81, 116, 132–134, 139, 140].

The aggregate cohesive strength τ is a function of the physicochemical conditions and floc properties, while the turbulent hydrodynamic stress σ depends on the design of the aggregation unit (geometry) and the mixing intensity [7, 75, 81, 117, 132, 134, 141, 142]. A number of empirical and theoretical models have been proposed for predicting the maximum hydrodynamic breaking force F_H and the global hydrodynamic stress σ acting on a spherical aggregate in the inertia and viscous domain of turbulence (Table 1.1).

He et al. [135], in a study of floc strength under turbulent flow conditions, derived a mathematical expression for an estimate of the floc binding or cohesive force F_B using a fractal dimension approach, while Lu et al. [15] in a similar study presented a theoretical model for the aggregate binding force F_B of spherical monodisperse particles, both of which are presented in Equations (1.21) and (1.22). Yuan and Farnood [143], as well as Jarvis et al. [144], in their review of aggregate strength and breakage gave an estimate of the floc strength τ and global hydrodynamic stress σ obtained from empirical studies of various types of flocs:

$$F_B = 48^{-\frac{2}{3}}\pi^{\frac{5}{3}}d^{\frac{2}{3}}\sigma(\rho_o-\rho_w)^{\frac{2}{3}}d^{(1+\frac{D_f}{3})} \tag{1.21}$$

$$F_B = \frac{(1-P_f)F_{H\max}}{P_f d_p^2} \tag{1.22}$$

The global hydrodynamic stress σ due to the shearing action of the fluid motion on the floc as well as the overall mechanical strength of an aggregate τ, assuming a uniform floc shape and constant porosity, can be expressed mathematically in Equations (1.23) and (1.24) [77, 116]. In turbulent hydrodynamic flow, the

TABLE 1.1 Empirical and Theoretical Models of Maximum Hydrodynamic Breaking Force

MODELS	*MAXIMUM HYDRODYNAMIC FORCE*	*REFERENCES*
1	$F_{H\max} = \rho\varepsilon^{\frac{2}{3}} d_f^{\frac{8}{3}}$	[27, 145]
2	$F_{H\max} = \frac{5}{8}\pi\mu d^2 G$	[140]
3	$F_{H\max} = \frac{5}{8}\rho C_1 \varepsilon^{\frac{2}{3}} d^{\frac{8}{3}}$	[59, 140]
4	$F_{H\max} = \left(\frac{\pi\rho_f \varepsilon^{\frac{2}{3}} d^{\frac{8}{3}}}{4}\right)$	[135]
5	$F_{H\max} = \frac{\pi\rho\varepsilon}{60\mu} d^4$	[135]
6	$F_{H\max} = \frac{\rho_{f\varepsilon d_f^4}}{\mu}$	[145]
7	$F_{H\max} = \mu G d_f^2$	[81]
8	$F_{H\max} = \frac{\pi\mu d_f^2 G}{2}$	[119]
9	$F_{H\max} = \frac{\pi\sigma d_f^2}{4}$	[64, 146]
10	$F_{H\max} = \rho^{\frac{1}{3}} \varepsilon^{\frac{2}{3}} d_f^{\frac{8}{3}}$	[7]
11	$F_{H\max} = \frac{\rho\varepsilon d_f^4}{\mu}$	[7]
12	$F_{H\max} = d_F d_p \nu G = d_F d_p \sqrt{\varepsilon\nu}$	[67]

fluctuating motions of the fluid are responsible for shear, tensile, and compressive stress (normal stress) on the flocs, depending on the direction of the fluctuating velocities acting on the aggregates [67, 77, 146, 147]. The exact type and magnitude of the prevailing stress causing aggregate disruption depend mainly on the relation between the floc size and the eddy size or radius, r_w [67, 148]:

$$\sigma = \mu G = \mu\sqrt{\frac{\varepsilon}{\nu}} \tag{1.23}$$

$$\tau = \frac{(1-p)F_B}{p d_p^2} \tag{1.24}$$

1.7.2 Fluid–Particle Interactions and Floc Stability

The size of aggregates varies from molecular dimensions to a range that is visible to the unaided eye, with the smaller sizes being associated with the primary particles of diameter d_p, while the largest size d_{Fmax} is determined by the balance of floc growth and rupture within the fluid [64, 139, 149, 150]. The floc growth and breakage are known to occur simultaneously—growing flocs are subjected to breakage while fragments of broken flocs undergo agglomeration until a levelling off of the floc sizes at steady state when the maximum stable size d_{Fmax} is attained [54, 139, 151, 152].

There is an increase in floc size as long as the hydrodynamic force due to turbulent shear is less than the floc cohesive or binding force, and after an extended period of time, an equilibrium floc size distribution is reached. In this case, either a continued particle or microflocs attachment to the larger flocs is prevented, or floc breakup kinetics balances the turbulence-induced collision [7, 67]. Increasing the shear rate (agitation or mixing speed) or continued shearing beyond the steady state leads to a sharp reduction in the mean floc size [7, 15, 34, 139, 153, 154]. The steady-state phase is regarded as the balance between floc growth and breakage under a given shear condition. In the viscous subrange, Lu et al. [15] expressed this phenomenon in terms of the kinetic equation of flocculation as presented in Equation (1.25). The quantities N_A, N_B, σ_s, and τ_s represent the change in the particle number concentration per unit volume for aggregation and breakage, the aggregate shear strength, and the shearing stress, respectively:

$$\frac{dN}{dt} = -\frac{2}{3} \propto_0 \left(\frac{\varepsilon}{\nu}\right)^{\frac{1}{2}} d_A^3 N_A^2 + \beta_1 \frac{\tau_s}{\sigma_s} \left(\frac{\varepsilon}{\nu}\right)^{\frac{1}{2}} d_B^2 N_B \tag{1.25}$$

In the analysis of floc breakup under turbulent hydrodynamic conditions, three approaches are normally employed, namely the limiting strength approach, the maximum strain rate approach, and the maximum floc size approach [6]. The first approach to floc breakup analysis, which is based on force balances [67], requires an accurate description of the flow field in order to describe the stress acting on the model floc. In this approach for analysing the breakage phenomenon, under a given agitation condition, the critical condition of floc breakage is attained when the hydrodynamic breaking force F_H is greater than or equal to the floc binding or cohesive force F_B ($B \leq 1$). Consequently, the conceptual form of the floc growth and breakage rate mechanism can be expressed mathematically in Equations (1.26) and (1.27) while Equations (1.28) and (1.29) express the critical condition of floc breakage for the inertia and viscous subrange, respectively [59, 59, 60, 64, 81, 116, 132–135, 140, 146]. A number of empirical and theoretical models have been developed for estimating the maximum aggregate size d_{Fmax}. A few of such expressions are presented for the inertia and viscous subrange of turbulence in Table 1.2 [143, 144]:

$$\beta_{floc} = \alpha_{ij} \beta_{col_{ij}} - \beta_{br} \tag{1.26}$$

$$\beta_{floc} = (\alpha_{ijBM} \beta_{BM} + \alpha_{ijSH} \beta_{SH} + \alpha_{ijDS} \beta_{DS}) - \beta_{br} \tag{1.27}$$

$$B = \frac{Floc\,binding\,force}{Hydrodynamic\,force} = \frac{F_B}{F_H} = \frac{48^{-\frac{2}{3}} \pi^{\frac{5}{3}} d^{\frac{2}{3}} \sigma (\rho_o - \rho_w)^{\frac{2}{3}} d^{(1+\frac{D_f}{3})}}{\dfrac{\pi \rho_w^{\frac{2}{3}} \mu^{\frac{4}{3}} G^{\frac{8}{3}} d^{\frac{8}{3}}}{4}} \tag{1.28}$$

$$B = \frac{Floc\,binding\,force}{Hydrodynamic\,force} = \frac{F_B}{F_H} = \frac{48^{\frac{2}{3}} \pi^{\frac{5}{3}} d^{\frac{2}{3}} \sigma (\rho_o - \rho_w)^{\frac{2}{3}} d^{(1+\frac{D_f}{3})}}{\dfrac{\pi \rho_w d^4 G^2}{60}} \tag{1.29}$$

TABLE 1.2 Empirical and Theoretical Models for Estimating Maximum Floc Size under Steady-State Conditions

MODELS	*MAXIMUM FLOC DIAMETER*	*REFERENCES*
1	$d_{F\max} = K'\varepsilon^{-1(1+D_p)}$	[49]
2	$d_{F\max} = 0.68\rho^{\frac{3}{5}} d_0^{\frac{3}{5}} \varepsilon^{\frac{11}{20}} \nu^{\frac{9}{20}} F_B^{\frac{3}{5}}$	[63]
3	$d_{F\max} = \frac{6U(1-p)^{\frac{2}{3}}}{\pi^2 KG\eta d_p^{\ 2}}$	[119]
4	$d_{F\max} = \frac{(F_B / d_p)}{\rho\sqrt{\varepsilon v}}$	[67, 118]
5	$d_{F\max} = \left[\frac{\left(\frac{F_B}{d_p}\right)d_p^k}{\rho G\nu}\right]^{\frac{1}{k+1}}$	[67]
5	$d_{F\max} = \left[\frac{(\frac{F_B}{d_p})d_p^k}{\sigma\sqrt{\varepsilon\nu}}\right]^{\frac{1}{k+1}}$	[67]
6	$d_{F\max} = \frac{\sqrt{\frac{\sigma}{\rho_f}}}{\sqrt{\frac{\varepsilon}{\nu}}}$	[77]
7	$d_{F\max} = \frac{(\frac{\sigma}{\nu})^{\frac{1}{3}}}{\frac{1}{\varepsilon}}$	[77]
8	$d_{F\max} = 2\rho^{-\frac{1}{2}}(\varepsilon\nu)^{-\frac{1}{4}} B_f^{\frac{1}{2}}$	[15]
9	$d_{F\max} = \left[\frac{B_f \nu^{\frac{3}{4}}}{1.9\rho\varepsilon^{\frac{11}{12}} d_p^2}\right]^{\frac{3}{5}}$	[15]

The fluctuating instantaneous fluid velocities acting parallel to the surface of the floc will induce a local shear stress σ_s on the aggregate in the viscous subrange reaching a maximum value when the floc size is roughly equal to the eddy scale (Kolmogorov micro scale) [118]. Similarly, fluctuating fluid velocities normal to the floc surface or dynamic pressure fluctuations acting on opposite sides of an aggregate will result in normal or bulk pressure stress σ_t (tensile or compressive) [53]. In addition, turbulent drag forces F_D acting on the surface of an aggregate, which originates from the local motion of the fluid relative to the motion of the aggregates, will result in instantaneous surface shear forces and shear stress, respectively [77]. It has been shown that for flocs in the inertial subrange of turbulence ($d_F > \lambda$), tensile stress will predominate causing wholesale fracture or fragmentation [15], while for those in the viscous subrange (d_F

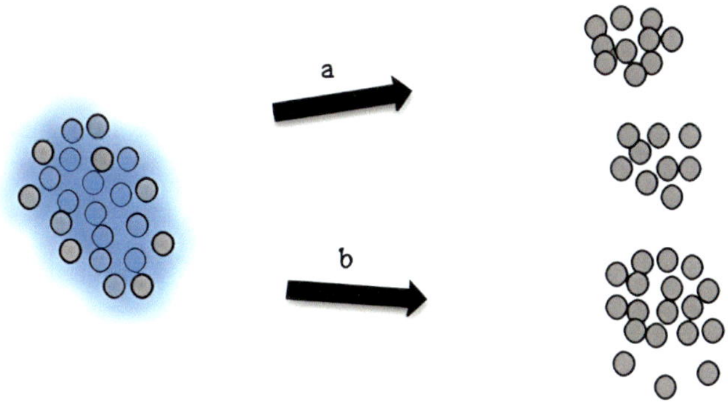

FIGURE 1.5 Schematic representation of hydrodynamic interactions leading to floc rupture: (a) floc splitting or fragmentation and (b) surface erosion.

$< \lambda$), shear stress will cause erosion of the particles, floc shell, or floc surface, as shown schematically in Figure 1.5 [47, 53].

1.7.3 Mechanisms of Aggregate Disruption

Several mechanisms have been proposed to account for the disruption of aggregates in orthokinetic flocculation [6–8, 46, 47, 57, 67, 117, 119, 144, 155, 156–158]. In addition to floc splitting, fracture or bulk rupture and surface erosion mentioned earlier, several other mechanisms have been proposed to be responsible for floc fragmentation but with little experimental data or theoretical analysis. Such mechanisms include aggregate–aggregate collisions, aggregate–stirrer collisions, as well as collisions with baffles and the tank wall [15, 138].

A number of theoretical and empirical models have been developed from micromechanical analysis to account for the hydrodynamic stress exerted on particle agglomerates and their cohesive strength. Considering a model floc undergoing rupture by fragmentation ($\sigma_t \geq \tau_t$) in the inertia domain of turbulence ($d_F > \lambda$), subjected to the shearing action of fluid motion, the turbulent hydrodynamic tensile stress σ_t causing bulk rupture or fragmentation by floc splitting and the aggregate tensile strength τ_t resisting the fluctuating pressure on the floc may be expressed mathematically in Equations (1.30) and (1.31):

$$\sigma_t = 0.49\rho_f(\varepsilon^3\nu)^{\frac{1}{4}}d_F \tag{1.30}$$

$$\tau_t = \frac{(1-\rho_f)B_f}{d_p^2} \tag{1.31}$$

Similarly, for a floc rupture due to the erosion of primary particles, micro flocs, or floc shell ($\sigma_s \geq \tau_s$) in the viscous subrange ($d_F < \lambda$); the turbulent hydrodynamic shear stress σ_s eroding the surface of the aggregates and the corresponding aggregate shear strength τ_s resisting the viscous shear forces can be expressed mathematically in Equations (1.32) and (1.33) [15]:

$$\sigma_s = 0.26\rho_f(\nu\varepsilon)^{\frac{1}{2}} \tag{1.32}$$

$$\tau_s = \frac{(1-\rho_f)B_f}{d_p^2} \tag{1.33}$$

In a related study, Mühle [67] presented floc breakup analysis on the basis of rheology. The strength of a model floc was described in terms of the surface shear yield strength τ_y resisting floc shell erosion due to pseudo-surface tension force γ_F eroding the particle chain of the outer floc surface in the viscous domain of turbulence ($\sigma_s \geq \tau_y$) (Equation 1.34). The yield stress approach was also presented by Liu et al. [147] for calculating the maximum floc tensile yield stress σ_y at which breakage is likely to occur in the inertia subrange ($\sigma_t \geq \sigma_y$) (Equation 1.35):

$$\tau_y = \frac{F_B}{d_p^2}\left(\frac{\lambda_0}{d_F}\right)^k \tag{1.34}$$

$$\sigma_y = 0.5\alpha\rho\varepsilon^{\frac{2}{3}}\left[\frac{2\pi}{d_f}\right]^{-\frac{2}{3}} \tag{1.35}$$

On a similar basis, Attia [118] presented models for predicting the critical fluid velocity V_c above which there will be floc breakage by estimating floc yield stress σ_y (tensile or compressive) resulting from dynamic pressure acting on the floc (Equations 1.36 and 1.37). The kinetic equations for fragmentation and surface erosion mechanisms are presented in Equations (1.38) and (1.39) based on the empirical results from flocculation in a stirred reactor. The quantities N_A, N_B, σ_s, and τ_s represent the change in the particle number concentration per unit volume for aggregation and breakage, the aggregate shear strength, and the shearing stress, respectively [15, 49]:

$$V_c = \sqrt{\frac{2\sigma_y}{\rho_f}} \tag{1.36}$$

$$\sigma_y = \frac{1}{2}\rho_f v^2 \tag{1.37}$$

$$-\frac{dN_A}{dt} = KN_A^2 - \beta N_B \tag{1.38}$$

$$\frac{dN}{dt} = \beta_1 \frac{\tau_s}{\sigma_s}\left(\frac{\varepsilon}{\nu}\right)^{\frac{1}{2}} d_F^2 N_F \tag{1.39}$$

1.8 PRACTICAL APPLICATIONS AND FUTURE OUTLOOK

Over the past decades, significant progress has been made in understanding the complex turbulent interactions that influence and drive the flocculation process. Improvement in modelling and simulation tools has led to the use of computational techniques in turbulent flow investigations. Nowadays, computational fluid dynamics provide comprehensive information for process engineers necessary to design, optimise, or retrofit various unit processes involving fluid flow as well as improving our understanding of flows in natural systems. This is extremely useful, especially in investigations involving turbulent flows in devices with complex geometries with several design parameters requiring optimisation without any need to perform the physical experiments [64, 136].

However, the numerical study of turbulent flow and flocculation process is still being done independently, and only a few studies [159–164] deal with the integration of these interdependent processes. This is partly driven by the evolution pattern of computational fluid modelling (CFD), which was specifically developed for other industries but is now widely adopted across many fields [64]. Therefore, future studies must strive to advance research in this direction by coupling the modelling of turbulent fluid flow and aggregation processes within the computational environment in order to build on the recent progress made in these fields [64]. This in turn will assist process engineers in developing a new generation of phase separation reactors more efficiently and improve our understanding and ability to accurately predict flow conditions in natural systems that are vital for many environmental processes.

1.9 CONCLUDING REMARKS

The flocculation process has been shown to be an effective phase separation technique in engineered unit processes and is an important driver of many environmental processes (sediment transport, contaminant migration, etc.) in natural systems. The importance of hydrodynamics and flow turbulence as the main drivers of this process has been highlighted. Although significant progress has been made in the understanding of physicochemical aspects of flocculation, turbulence remains a complex and poorly understood phenomenon despite its profound influence on many engineered and life processes. This comprehensive study highlights the need for a closer integration of modelling and simulation tools in order to improve our understanding and ability to influence and predict these processes much more accurately.

DATA AVAILABILITY

Data sharing is not applicable to this chapter as no new data were created or analysed in this study.

REFERENCES

1. J. Gregory, Fundamentals of flocculation, *Crit. Rev. Environ. Control* 19 (1989) 185–230. https://doi.org/10.1080/10643388909388365.
2. I. Popa, G. Papastavrou, M. Borkovec, Charge regulation effects on electrostatic patch-charge attraction induced by adsorbed dendrimers, *Phys. Chem. Chem. Phys.* 12 (2010) 4863–4871. https://doi.org/10.1039/B925812D.
3. J. Gregory, The role of colloid interactions in solid–liquid separation, *Water Sci. Technol.* 27 (1993) 1–17.
4. J. Bratby, *Coagulation and Flocculation in Water and Wastewater Treatment*, IWA Publishing, London, 2006.
5. D.H. Bache, R. Gregory, *Flocs in Water Treatment*, IWA Publishing, London, 2007.
6. M.M. Benjamin, D.F. Lawler, *Water Quality Engineering: Physical/Chemical Treatment Processes*, John Wiley & Sons, Hoboken, NJ, 2013.
7. E. Partheniades, *Cohesive Sediments in Open Channels: Properties, Transport and Applications*, Butterworth-Heinemann, Oxford, 2009.
8. J. Gregory, *Particles in Water: Properties and Processes*, CRC Press, Boca Raton, FL, 2006.
9. N.K. Shammas, Coagulation and flocculation, in: L.K. Wang, Y.-T. Hung, N.K. Shammas (eds.), *Physicochemical Treatment Processes*, Humana Press, Totowa, NJ, 2005: pp. 103–139. https://doi.org/10.1385/1-59259-820-x:103.

10. J.S. Marshall, S. Li, *Adhesive Particle Flow: A Discrete-Element Approach*, Cambridge University Press, New York, NY, 2014.
11. N.I. Lebovka, Aggregation of charged colloidal particles, in: M. Müller (ed.), *Polyelectrolyte Complexes in the Dispersed and Solid State I*, Springer, Heidelberg, 2013: pp. 57–96. https://doi.org/10.1007/12_2012_171.
12. I. Nopens, *Modelling the Activated Sludge Flocculation Process: A Population Balance Approach*, PhD thesis, University of Ghent, 2005.
13. G. Moody, P. Norman, Chemical pre-treatment, in: S. Tarleton, R. Wakeman (eds.), *Solid–Liquid Separation: Scale-Up of Industrial Equipment*, Elsevier, Oxford, 2005: pp. 38–81.
14. J.S. Laskowski, R.J. Pugh, Dispersion stability and dispersing agents, in: J.S. Laskowski, J. Ralston (eds.), *Colloid Chemistry in Mineral Processing*, Elsevier, Amsterdam, 1992: pp. 115–170. https://doi.org/10.1016/B978-0-444-88284-4.50009-3.
15. S. Lu, Y. Ding, J. Guo, Kinetics of fine particle aggregation in turbulence, *Adv. Colloid Interface Sci.* 78 (1998) 197–235. https://doi.org/10.1016/S0001-8686(98)00062-1.
16. K.J. Wilkinson, A. Reinhardt, Contrasting roles of natural organic matter on colloidal stabilization and flocculation, in: S.N. Liss, I.G. Droppo, G.G. Leppard, T.G. Milligan (eds.), *Flocculation in Natural and Engineered Environmental Systems*, CRC Press, Boca Raton, FL, 2005: pp. 143–170. https://doi.org/10.1201/9780203485330.
17. J. Gregory, *Particles in Water: Properties and Processes*, CRC Press, Boca Raton, FL, 2006. https://doi.org/10.1201/9780203508459.
18. T. Smith-Palmer, R. Pelton, Flocculation of particles, in: P. Somasundaran (ed.), *Encyclopedia of Surface and Colloid Science*, 2nd ed., CRC Press, Boca Raton, FL, 2006: pp. 2584–2599. https://doi.org/10.1081/E-ESCS3-120000076.
19. L.L. Schramm, *Emulsions, Foams, and Suspensions*, Wiley-VCH, Weinheim, 2005.
20. J. Addai-Mensah, C.A. Prestidge, Structure formation in dispersed systems, in: B. Dobias, H. Stechemesser (eds.), *Coagulation and Flocculation*, 2nd ed., CRC Press, Boca Raton, FL, 2005: pp. 135–216. https://doi.org/10.1201/9781420027686.
21. J. Gregory, Stability and flocculation of suspensions, in: A.P. Shamlou (ed.), *Processing of Solid–Liquid Suspensions*, Butterworth-Heinemann, Oxford, 1993: pp. 59–92. https://doi.org/10.1016/B978-0-7506-1134-3.50007-7.
22. D. Grasso, K. Subramaniam, M. Butkus, K. Strevett, J. Bergendahl, A review of non-DLVO interactions in environmental colloidal systems, *Rev. Environ. Sci. Biotechnol.* 1 (2002) 17–38. https://doi.org/10.1023/A:1015146710500.
23. J. Gregory, Flocculation of fine particles, in: P. Mavros, K.A. Matis (eds.), *Innovations in Flotation Technology*, Springer Netherlands, Heidelberg, 1992: pp. 101–124. https://doi.org/10.1007/978-94-011-2658-8_4.
24. P. Taboada-Serrano, C.-J. Chin, S. Yiacoumi, C. Tsouris, Modeling aggregation of colloidal particles, *Curr. Opin. Colloid Interface Sci.* 10 (2005) 123–132. https://doi.org/10.1016/j.cocis.2005.07.003.
25. A.T. Hanson, J.L. Cleasby, The effects of temperature on turbulent flocculation: Fluid dynamics and chemistry, *J. Am. Water Works Assoc.* 82 (1990) 56–73.
26. W. Pietsch, *Agglomeration Processes: Phenomena, Technologies, Equipment*, Wiley-VCH, Weinheim, 2002.
27. R.S. Farinato, S.-Y. Huang, P. Hawkins, Polyelectrolyte-assisted dewatering, in: R.S. Farinato, P.L. Dubin (eds.), *Colloid–Polymer Interactions: From Fundamentals to Practice*, John Wiley & Sons, New York, NY, 1993: pp. 3–50.
28. W. Lick, *Sediment and Contaminant Transport in Surface Waters*, CRC Press, Boca Raton, FL, 2008.
29. W. Lick, H. Huang, R. Jepsen, Flocculation of fine-grained sediments due to differential settling, *J. Geophys. Res. Oceans* 98 (1993) 10279–10288. https://doi.org/10.1029/93JC00519.
30. V. Runkana, P. Somasundaran, P.C. Kapur, A population balance model for flocculation of colloidal suspensions by polymer bridging, *Chem. Eng. Sci.* 61 (2006) 182–191. https://doi.org/10.1016/j.ces.2005.01.046.
31. O.O. Prat, J.J. Ducoste, Simulation of flocculation in stirred vessels: Lagrangian versus Eulerian, *Chem. Eng. Res. Des.* 85 (2007) 207–219. https://doi.org/10.1205/cherd05001.
32. O.P. Prat, J.J. Ducoste, Modeling spatial distribution of floc size in turbulent processes using the quadrature method of moments and computational fluid dynamics, *Chem. Eng. Sci.* 61 (2006) 75–86. https://doi.org/10.1016/j.ces.2004.11.070.
33. S.J. Curran, R.A. Black, Taylor-vortex bioreactors for enhanced mass transport, in: J. Chaudhuri, M. Al-Rubeai (eds.), *Bioreactors for Tissue Engineering: Principles, Design and Operation*, Springer, Dordrecht, 2005: pp. 47–85. https://doi.org/10.1007/1-4020-3741-4_3.
34. W. Wu, *Computational River Dynamics*, CRC Press, London, 2008.

35. M. Sievers, S.M. Stoll, C. Schroeder, M. Niedermeiser, T.I. Onyeche, Sludge dewatering and aggregate formation effects through Taylor vortex assisted flocculation, *Sep. Sci. Technol.* 43 (2008) 1595–1609. https://doi.org/10.1080/01496390801973888.
36. P.F. Tooby, G.L. Wick, J.D. Isaacs, The motion of a small sphere in a rotating velocity field: A possible mechanism for suspending particles in turbulence, *J. Geophys. Res.* 82 (1977) 2096–2100. https://doi.org/10.1029/JC082i015p02096.
37. S. Biggs, Aggregate structures and solid–liquid separation processes, *KONA Powder Part. J.* 24 (2006) 41–53.
38. J. Gregory, L. Guibai, Effects of dosing and mixing conditions on polymer flocculation of concentrated suspensions, *Chem. Eng. Commun.* 108 (1991) 3–21. https://doi.org/10.1080/00986449108910948.
39. M.A. Yukselen, J. Gregory, The effect of rapid mixing on the break-up and re-formation of flocs, *J. Chem. Technol. Biotechnol.* 79 (2004) 782–788. https://doi.org/10.1002/jctb.1056.
40. K.E. Lee, N. Morad, T.T. Teng, B.T. Poh, Development, characterization and the application of hybrid materials in coagulation/flocculation of wastewater: A review, *Chem. Eng. J.* 203 (2012) 370–386. https://doi.org/10.1016/j.cej.2012.06.109.
41. M. Hjorth, M.L. Christensen, Evaluation of methods to determine flocculation procedure for manure separation, *Trans. ASABE* 51 (2008) 2093–2103. https://doi.org/10.13031/2013.25391.
42. B.E. Logan, *Environmental Transport Processes*, John Wiley & Sons, Hoboken, NJ, 2012.
43. T.G. Milligan, P.S. Hill, A laboratory assessment of the relative importance of turbulence, particle composition, and concentration in limiting maximal floc size and settling behaviour, *J. Sea Res.* 39 (1998) 227–241. https://doi.org/10.1016/S1385-1101(97)00062-2.
44. E. Kissa, *Dispersions: Characterization, Testing, and Measurement*, Marcel Dekker, New York, NY, 1999.
45. J. Gregory, Flocculation fundamentals, in: T. Tadros (ed.), *Encyclopedia of Colloid and Interface Science*, Springer, Heidelberg, 2013: pp. 459–491. https://doi.org/10.1007/978-3-642-20665-8_17.
46. W. Van Leussen, Aggregation of particles, settling velocity of mud flocs – A review, in: J. Dronkers, W. Van Leussen (eds.), *Physical Processes in Estuarine*, Springer, Heidelberg, 2011: pp. 347–403. https://doi.org/10.1007/978-3-642-73691-9_19.
47. D.N. Thomas, S.J. Judd, N. Fawcett, Flocculation modelling: A review, *Water Res.* 33 (1999) 1579–1592. https://doi.org/10.1016/S0043-1354(98)00392-3.
48. J.F. Atkinson, R.K. Chakraborti, J.E. Benschoten, Effects of floc size and shape in particle aggregation, in: S.N. Liss, I.G. Droppo, G.G. Leppard, T.G. Milligan (eds.), *Flocculation in Natural and Engineered Environmental Systems*, CRC Press, Boca Raton, FL, 2005: pp. 95–120. https://doi.org/10.1201/9780203485330.
49. T.A. Kramer, M.M. Clark, Incorporation of aggregate breakup in the simulation of orthokinetic coagulation, *J. Colloid Interface Sci.* 216 (1999) 116–126. https://doi.org/10.1006/jcis.1999.6305.
50. W. Lick, J. Lick, C.K. Ziegler, Flocculation and its effect on the vertical transport of fine-grained sediments, in: B.T. Hart, P.G. Sly (eds.), *Sediment/Water Interactions*, Springer, Heidelberg, 1992: pp. 1–16. https://doi.org/10.1007/978-94-011-2783-7_1.
51. W. Lick, J. Lick, C.K. Ziegler, Flocculation and its effect on the vertical transport of fine-grained sediments, *Hydrobiologia* 235–236 (1992) 1–16. https://doi.org/10.1007/BF00026196
52. F.D. Lawler, Physical aspects of flocculation: From microscale to macroscale, *Water Res.* 27 (1993) 165–180.
53. K.A. Kruster, *The Influence of Turbulence on Aggregation of Small Particles in Agitated Vessels*, PhD thesis, Technische Universiteit Eindhoven, 1991.
54. W. Lick, J. Lick, Aggregation and disaggregation of fine-grained lake sediments, *J. Great Lakes Res.* 14 (1988) 514–523. https://doi.org/10.1016/S0380-1330(88)71583-X.
55. C.-H. Tsai, S. Iacobellis, W. Lick, Flocculation of fine-grained lake sediments due to a uniform shear stress, *J. Great Lakes Res.* 13 (1987) 135–146. https://doi.org/10.1016/S0380-1330(87)71637-2.
56. L. Wang, D.L. Marchisio, R.D. Vigil, R.O. Fox, CFD simulation of aggregation and breakage processes in laminar Taylor–Couette flow, *J. Colloid Interface Sci.* 282 (2005) 380–396. https://doi.org/10.1016/j.jcis.2004.08.127.
57. J. Gregory, Floc formation and floc structure, in: G. Newcombe, D. Dixon (eds.), *Interface Science in Drinking Water Treatment: Theory and Applications*, Academic Press, London, 2006: pp. 25–43. https://doi.org/10.1016/S1573-4285(06)80072-4.
58. R.D. Letterman, A. Amirtharajah, C.R. O'Melia, Coagulation and flocculation, in: J. Edzwald (ed.), *Water Quality & Treatment: A Handbook on Drinking Water*, McGraw-Hill, New York, NY, 2010: pp. 6.1–6.66.
59. J. Bridgeman, B. Jefferson, S.A. Parsons, The development and application of CFD models for water treatment flocculators, *Adv. Eng. Softw.* (2010). https://doi.org/10.1016/j.advengsoft.2008.12.007.
60. J. Bridgeman, B. Jefferson, S. Parsons, Assessing floc strength using CFD to improve organics removal, *Chem. Eng. Res. Des.* 86 (2008) 941–950. https://doi.org/10.1016/j.cherd.2008.02.007.

61. T.R. Camp, P.C. Stein, Velocity gradients and internal work in fluid motion, *J. Boston Soc. Civ. Eng.* 30 (1943) 219–237.
62. J.C. Winterwerp, A simple model for turbulence-induced flocculation of cohesive sediment, *J. Hydraul. Res.* 36 (1998) 309–326. https://doi.org/10.1080/00221689809498621.
63. Z. Zhu, Theory on orthokinetic flocculation of cohesive sediment: A review, *J. Geosci. Environ. Prot.* 2 (2014). https://doi.org/10.4236/gep.2014.25003.
64. J. Bridgeman, B. Jefferson, S.A. Parsons, Computational fluid dynamics modelling of flocculation in water treatment: A review, *Eng. Appl. Comput. Fluid Mech.* 3 (2009) 220–241. https://doi.org/10.1080/19942060.2009.11015267.
65. J. Korpijärvi, E. Laine, H. Ahlstedt, Using CFD in the study of mixing in coagulation and flocculation, in: H.H. Hahn, E. Hoffmann, H. Odegaard (eds.), *Chemical Water and Wastewater Treatment VI*, Springer, Heidelberg, 2000: pp. 89–99. https://doi.org/10.1007/978-3-642-59791-6_9.
66. T.A. Kramer, M.M. Clark, Influence of strain-rate on coagulation kinetics, *J. Environ. Eng.* 123 (1997) 444–452. https://doi.org/10.1061/(ASCE)0733-9372(1997)123:5(444).
67. K. Mühle, Floc stability in laminar and turbulent flow, in: B. Dobias (ed.), *Coagulation and Flocculation: Theory and Applications*, Marcel Dekker, New York, NY, 1993: pp. 355–390.
68. L. Svarovsky, *Solid–Liquid Separation*, 4th ed., Butterworth-Heinemann, Woburn, MA, 2000.
69. K.J. Ives, Experiments in orthokinetic flocculation, in: J. Gregory (ed.), *Solid–Liquid Separation*, Ellis Horwood Ltd, London, 1984: pp. 196–220.
70. G. Belfort, Fluid mechanics and cross-flow membrane filtration, in: H.S. Muralidhara (ed.), *Advances in Solid–Liquid Separation*, Battelle Press, Columbus, OH, 1986: pp. 165–189.
71. P.T. Spicer, *Shear-Induced Aggregation-Fragmentation: Mixing and Aggregate Morphology Effects*, PhD thesis, University of Cincinnati, 1997.
72. L. Falk, J.-M. Commenge, Characterization of mixing and segregation in homogeneous flow systems, in: V. Hessel, A. Renken, J.C. Schouten, J.-I. Yoshida (eds.), *Handbook of Micro Reactors*, John Wiley & Sons, Weinheim, 2009: pp. 147–171. https://doi.org/10.1002/9783527631445.ch6.
73. F. Concha, *Solid-Liquid Separation in the Mining Industry*, Springer, Heidelberg, 2014.
74. J.B. Farrow, J.D. Swift, A new procedure for assessing the performance of flocculants, *Int. J. Miner. Process.* 46 (1996) 263–275. https://doi.org/10.1016/0301-7516(95)00084-4.
75. E. Carissimi, J. Rubio, Polymer-bridging flocculation performance using turbulent pipe flow, *Miner. Eng.* 70 (2015) 20–25. https://doi.org/10.1016/j.mineng.2014.08.019.
76. D.W. Hendricks, *Fundamentals of Water Treatment Unit Processes: Physical, Chemical, and Biological*, CRC Press, Boca Raton, FL, 2011.
77. A.P. Shamlou, N. Hooker-Titchener, Turbulent aggregation and breakup of particles in liquids in stirred vessels, in: A.P. Shamlou (ed.), *Processing of Solid-Liquid Suspensions*, Butterworth-Heinemann, Oxford, 1993: pp. 1–25. https://doi.org/10.1016/B978-0-7506-1134-3.50005-3.
78. R. Hogg, Flocculation and dewatering, *Int. J. Miner. Process.* 58 (2000) 223–236. https://doi.org/10.1016/S0301-7516(99)00023-X.
79. B. Bergenstahl, Emulsions, in: S.T. Beckett (ed.), *Physico-Chemical Aspects of Food Processing*, Blackie Academic & Professional, Glasgow, 1995: pp. 49–64. https://doi.org/10.1007/978-1-4613-1227-7_3.
80. M. Son, T.-J. Hsu, Flocculation model of cohesive sediment using variable fractal dimension, *Environ. Fluid Mech.* 8 (2008) 55–71. https://doi.org/10.1007/s10652-007-9050-7.
81. Y. Adachi, A. Kobayashi, M. Kobayashi, Structure of colloidal flocs in relation to the dynamic properties of unstable suspension, *Int. J. Polym. Sci.* 2012 (2012) 1–14. https://doi.org/10.1155/2012/574878.
82. N. Tambo, Optimization of flocculation in connection with various solid-liquid separation processes, in: H. Hahn, R. Klute (eds.), *Chemical Water and Wastewater Treatment*, Springer, Heidelberg, 1990: pp. 17–32. https://doi.org/10.1007/978-3-642-76093-8_2.
83. M. Yusa, H. Suzuki, S. Tanaka, Separating liquids from solids by pellet flocculation, *J. Am. Water Works Assoc.* 67 (1975) 397–402.
84. M. Yusa, C. Igarashi, Compaction of flocculated material, *Water Res.* 18 (1984) 811–816. https://doi.org/10.1016/0043-1354(84)90264-1.
85. K. Higashitani, T. Shibata, Y. Matsuno, Formation of pellet flocs from kaolin suspension and their properties, *J. Chem. Eng. Jpn.* 20 (1987) 152–157.
86. M. Yusa, A.M. Gaudin, Formation of pellet-like flocs of kaolinite by polymer chains, *Am. Ceram. Soc. Bull.* 43 (1964) 402–406.
87. M. Yusa, Mechanisms of pelleting flocculation, *Int. J. Miner. Process.* 4 (1977) 293–305. https://doi.org/10.1016/0301-7516(77)90010-2.

88. X.C. Wang, P.K. Jin, H.L. Yuan, E.R. Wang, N. Tambo, Pilot study of a fluidized-pellet-bed technique for simultaneous solid/liquid separation and sludge thickening in a sewage treatment plant, *Water Sci. Technol.* 49 (2004) 81–88.
89. Z. Gang, H. Ting-lin, T. Chi, Z. Li, H. Wen-jie, H. Hong-da, L. Chen, Settling behaviour of pellet flocs in pelleting flocculation process: Analysis through operational conditions, *Water Sci. Technol.* 62 (2010) 1346–1352. https://doi.org/10.2166/wst.2010.429.
90. S. Bähr, *Experimental Studies of Fundamental Processes of Pelleting Flocculation*, PhD thesis, Brandenburg University of Technology, 2006.
91. W. Walaszek, *Investigation Upon Structure of Pellet Flocs against Process Performance as a Tool to Optimize Sludge Conditioning*, PhD thesis, Brandenburg University of Technology, 2007.
92. T. Panswad, S. Polwanich, Pilot plant application of pelletisation process on low-turbidity river water, *J. Water Supply: Res. Technol. AQUA* 47 (1998) 236–244.
93. L. Glasgow, *Physicochemical Influences Upon Floc Deformability, Density, and Permeability*, Glasgow, Scotland, 2005: pp. 1–10.
94. L. Gillberg, B. Hanse, I. Karlsson, A. Enkel, A. Palsson, *About Water Treatment*, Kemira Kemwater, Helsingborg, 2003.
95. M. Yusa, Pelleting flocculation in sludge conditioning—An overview, in: Y.A. Attia (ed.), *Flocculation in Biotechnology and Separation Systems*, Elsevier, Amsterdam, 1987: pp. 755–763.
96. A. Hemme, R. Polte, P. Ay, Pelleting flocculation—The alternative to traditional sludge conditioning, *Aufbereitungs-Technik* 36 (1995) 226–235.
97. A. Amirtharajah, N. Tambo, Mixing in water treatment, in: A. Amirtharajah, M.M. Clark, R. Trussell (eds.), *Mixing in Coagulation and Flocculation*, American Water Works Association, Denver, CO, 1991: pp. 3–34.
98. K. Higashitani, T. Kubota, Pelleting flocculation of colloidal latex particles, *Powder Technol.* 51 (1987) 61–69. https://doi.org/10.1016/0032-5910(87)80040-2.
99. V.E. Vigdergauz, G. Yu. Gol'berg, Kinetics of mechanical floccule synaeresis, *J. Min. Sci.* 48 (2012) 347–353. https://doi.org/10.1134/S1062739148020165.
100. W. Walaszek, P. Ay, Extended interpretation of the structural attributes of pellet flocs in pelleting flocculation, *Miner. Eng.* 19 (2006) 1397–1400. https://doi.org/10.1016/j.mineng.2006.03.004.
101. W. Walaszek, P. Ay, Porosity and interior structure analysis of pellet-flocs, *Colloids Surf. A: Physicochem. Eng. Asp.* 280 (2006) 155–162. https://doi.org/10.1016/j.colsurfa.2006.01.049.
102. W. Walaszek, P. Ay, Pelleting flocculation—An alternative technique to optimise sludge conditioning, *Int. J. Miner. Process.* 76 (2005) 173–180. https://doi.org/10.1016/j.minpro.2005.01.001.
103. N. Tambo, C.C. Wang, The mechanism of pellet flocculation in fluidized-bed operations, *J. Water Supply: Res. Technol. AQUA* 42 (1993) 67–76.
104. M. Hjorth, *Flocculation and Solid-Liquid Separation of Animal Slurry: Fundamentals, Control and Application*, PhD thesis, University of Southern Denmark, 2009.
105. X.H. Wang, C. Jiang, Papermaking Part II: Surface and colloid chemistry of papermaking process, in: P. Somasundaran (ed.), *Encyclopedia of Surface and Colloid Science*, 2nd ed., CRC Press, Boca Raton, FL, 2006: pp. 4435–4451. https://doi.org/10.1081/E-ESCS3 120028090.
106. H. Xiao, Fine clay flocculation, in: P. Somasundaran (ed.), *Encyclopedia of Surface and Colloid Science*, 2nd ed., CRC Press, Boca Raton, FL, 2006: pp. 2572–2583. https://doi.org/10.1081/E-ESCS3.
107. G. Petzold, S. Schwarz, Polyelectrolyte complexes in flocculation applications, in: M. Müller (ed.), *Polyelectrolyte Complexes in the Dispersed and Solid State II*, Springer, Heidelberg, 2013: pp. 25–65. https://doi.org/10.1007/12_2012_205.
108. B.M. Moudgil, Selection of flocculants for solid-liquid separation process, in: H.S. Muralidhara (ed.), *Advances in Solid-Liquid Separation*, Battelle Press, Columbus, OH, 1986: pp. 191–204.
109. N. Böhm, W.-M. Kulicke, Optimization of the use of polyelectrolytes for dewatering industrial sludges of various origins, *Colloid Polym. Sci.* 275 (1997) 73–81. https://doi.org/10.1007/s003960050054.
110. L. Besra, D.K. Sengupta, S.K. Roy, P. Ay, Polymer adsorption: Its correlation with flocculation and dewatering of kaolin suspension in the presence and absence of surfactants, *Int. J. Miner. Process.* 66 (2002) 183–202. https://doi.org/10.1016/S0301-7516(02)00064-9.
111. M. Hjorth, M.L. Christensen, P.V. Christensen, Flocculation, coagulation, and precipitation of manure affecting three separation techniques, *Bioresour. Technol.* 99 (2008) 8598–8604. https://doi.org/10.1016/j.biortech.2008.04.009.
112. M. Hjorth, B.U. Jørgensen, Polymer flocculation mechanism in animal slurry established by charge neutralization, *Water Res.* 46 (2012) 1045–1051. https://doi.org/10.1016/j.watres.2011.11.078.
113. C.H. Lee, J.C. Liu, Enhanced sludge dewatering by dual polyelectrolytes conditioning, *Water Res.* 34 (2000) 4430–4436. https://doi.org/10.1016/S0043-1354(00)00209-8.

114. G. Lagaly, From clay mineral crystals to colloidal clay mineral dispersions, in: B. Dobias (ed.), *Coagulation and Flocculation: Theory and Applications*, Marcel Dekker, New York, NY, 1993: pp. 427–494.
115. G. Lagaly, From clay mineral crystals to colloidal clay mineral dispersions, in: B. Dobias, H. Stechemesser (eds.), *Coagulation and Flocculation: Second Edition*, 2nd ed., CRC Press, Boca Raton, FL, 2005: pp. 519–600. https://doi.org/10.1201/9781420027686.
116. C. Coufort, D. Bouyer, A. Liné, Flocculation related to local hydrodynamics in a Taylor–Couette reactor and in a jar, *Chem. Eng. Sci.* 60 (2005) 2179–2192. https://doi.org/10.1016/j.ces.2004.10.038.
117. J.F. Boyle, I. Manas-Zloczower, D.L. Feke, Hydrodynamic analysis of the mechanisms of agglomerate dispersion, *Powder Technol.* 153 (2005) 127–133. https://doi.org/10.1016/j.powtec.2004.08.010.
118. Y.A. Attia, Flocculation, in: J.S. Laskowski, J. Ralston (eds.), *Colloid Chemistry in Mineral Processing*, Elsevier, Amsterdam, 1992: pp. 277–308. https://doi.org/10.1016/B978-0-444-88284-4.50014-7.
119. N.N. Rulyov, Physicochemical microhydrodynamics of ultradisperse systems, in: V.M. Starov (ed.), *Nanoscience: Colloidal and Interfacial Aspects*, CRC Press, Boca Raton, FL, 2010: pp. 969–995. https://doi.org/10.1201/EBK1420065008.
120. M. Zlokarnik, *Stirring: Theory and Practice*, Wiley-VCH, Weinheim, 2008.
121. J. Baldyga, J.R. Bourne, A fluid mechanical approach to turbulent mixing and chemical reaction Part II: Micromixing in the light of turbulence theory, *Chem. Eng. Commun.* 28 (1984) 243–258. https://doi.org/10.1080/00986448408940136.
122. S.F. Thomas, P. Rooks, F. Rudin, N. Cagney, S. Balabani, S. Atkinson, P. Goddard, R.M. Bransgrove, P.T. Mason, M.J. Allen, Swirl flow bioreactor containing dendritic copper-containing alginate beads: A potential rapid method for the eradication of *Escherichia coli* from waste water streams, *J. Water Process Eng.* 5 (2015) 6–14. https://doi.org/10.1016/j.jwpe.2014.10.010.
123. S.M. Kresta, R.S. Brodkey, Turbulence in mixing applications, in: E.L. Paul, V.A. Atiemo-Obeng, S.M. Kresta (eds.), *Handbook of Industrial Mixing: Science and Practice*, John Wiley & Sons, Hoboken, NJ, 2004: pp. 19–87. https://doi.org/10.1002/0471451452.ch2.
124. D. Thoenes, *Chemical Reactor Development: From Laboratory Synthesis to Industrial Production*, Springer, Dordrecht, 1998.
125. T. Sparks, *Fluid Mixing in Rotor/Stator Mixers*, PhD thesis, Cranfield University, 1996.
126. J. Bałdyga, R. Pohorecki, Turbulent micromixing in chemical reactors — A review, *Chem. Eng. J. Biochem. Eng. J.* 58 (1995) 183–195. https://doi.org/10.1016/0923-0467(95)02982-6.
127. J.Y. Oldshue, R.R. Trussell, Design of impellers for mixing, in: A. Amirtharajah, M.M. Clark, R. Trussell (eds.), *Mixing in Coagulation and Flocculation*, American Water Works Association, Denver, CO, 1991: pp. 309–342.
128. N. Kockmann, *Transport Phenomena in Micro Process Engineering*, Springer, Heidelberg, 2008.
129. F. Maggi, *Flocculation Dynamics of Cohesive Sediment*, PhD thesis, Delft University of Technology, 2005.
130. J. Bałdyga, J.R. Bourne, *Turbulent Mixing and Chemical Reactions*, Wiley-VCH, Weinheim, 1999.
131. H. Wu, G.K. Patterson, Laser-Doppler measurements of turbulent-flow parameters in a stirred mixer, *Chem. Eng. Sci.* 44 (1989) 2207–2221. https://doi.org/10.1016/0009-2509(89)85155-3.
132. M. Kobayashi, Y. Adachi, S. Ooi, Breakup of fractal flocs in a turbulent flow, *Langmuir* 15 (1999) 4351–4356. https://doi.org/10.1021/la980763o.
133. D. Bouyer, A. Liné, Z. Do-Quang, Experimental analysis of floc size distribution under different hydrodynamics in a mixing tank, *AIChE J.* 50 (2004) 2064–2081. https://doi.org/10.1002/aic.10242.
134. D. Bouyer, C. Coufort, A. Liné, Z. Do-Quang, Experimental analysis of floc size distributions in a 1-L jar under different hydrodynamics and physicochemical conditions, *J. Colloid Interface Sci.* 292 (2005) 413–428. https://doi.org/10.1016/j.jcis.2005.06.011.
135. J. He, J. Liu, Y. Yuan, J. Zhang, A novel quantitative method for evaluating floc strength under turbulent flow conditions, *Desal. Water Treat.* (2014). https://doi.org/10.1080/19443994.2014.958107.
136. C.D. Argyropoulos, N.C. Markatos, Recent advances on the numerical modelling of turbulent flows, *Appl. Math. Model.* 39 (2015) 693–732. https://doi.org/10.1016/j.apm.2014.07.001.
137. P. Bubakova, M. Pivokonsky, P. Filip, Effect of shear rate on aggregate size and structure in the process of aggregation and at steady state, *Powder Technol.* 235 (2013) 540–549. https://doi.org/10.1016/j.powtec.2012.11.014.
138. G.G. Bemmer, *Agglomeration in Suspension: A Study of Mechanisms and Kinetics*, PhD thesis, Delft University of Technology, 1979.
139. P.T. Spicer, S.E. Pratsinis, Shear-induced flocculation: The evolution of floc structure and the shape of the size distribution at steady state, *Water Res.* 30 (1996) 1049–1056. https://doi.org/10.1016/0043-1354(95)00253-7.
140. M. Soos, A.S. Moussa, L. Ehrl, J. Sefcik, H. Wu, M. Morbidelli, Effect of shear rate on aggregate size and morphology investigated under turbulent conditions in stirred tank, *J. Colloid Interface Sci.* 319 (2008) 577–589. https://doi.org/10.1016/j.jcis.2007.12.005.

141. E. Carissimi, J. Rubio, The Flocs Generator Reactor–FGR: A new basis for flocculation and solid–liquid separation, *Int. J. Mineral Process.* 75 (2005) 237–247. https://doi.org/10.1016/j.minpro.2004.08.021.
142. E. Carissimi, J.D. Miller, J. Rubio, Characterization of the high kinetic energy dissipation of the Flocs Generator Reactor (FGR), *Int. J. Mineral Process.* 85 (2007) 41–49. https://doi.org/10.1016/j.minpro.2007.08.001.
143. Y. Yuan, R.R. Farnood, Strength and breakage of activated sludge flocs, *Powder Technol.* 199 (2010) 111–119. https://doi.org/10.1016/j.powtec.2009.11.021.
144. P. Jarvis, B. Jefferson, J. Gregory, S.A. Parsons, A review of floc strength and breakage, *Water Res.* 39 (2005) 3121–3137. https://doi.org/10.1016/j.watres.2005.05.022.
145. N. Tambo, R.J. François, Mixing, breakup and floc characteristics, in: A. Amirtharajah, M.M. Clark, R. Trussell (eds.), *Mixing in Coagulation and Flocculation*, American Water Works Association, Denver, CO, 1991: pp. 256–281.
146. S.X. Liu, L.A. Glasgow, Aggregate disintegration in turbulent jets, *Water Air Soil Pollut.* 95 (1997) 257–275. https://doi.org/10.1007/BF02406169.
147. L.A. Glasgow, X. Liu, Response of aggregate structures to hydrodynamic stress, *AIChE J.* 37 (1991) 1411–1414. https://doi.org/10.1002/aic.690370913.
148. G. Wang, S. Zhou, J.B. Joshi, G.J. Jameson, G.M. Evans, An energy model on particle detachment in the turbulent field, *Miner. Eng.* 69 (2014) 165–169. https://doi.org/10.1016/j.mineng.2014.07.018.
149. D.H. Bache, Floc rupture and turbulence: A framework for analysis, *Chem. Eng. Sci.* 59 (2004) 2521–2534. https://doi.org/10.1016/j.ces.2004.01.055.
150. E. Partheniades, Turbulence, flocculation and cohesive sediment dynamics, in: A.J. Mehta (ed.), *Nearshore and Estuarine Cohesive Sediment Transport*, American Geophysical Union, Washington, DC, 1993: pp. 40–59. https://doi.org/10.1029/CE042p0040.
151. R. Hogg, Flocculation and dewatering of fine-particle suspension, in: B. Dobias, H. Stechemesser (eds.), *Coagulation and Flocculation: Second Edition*, 2nd ed., CRC Press, Boca Raton, FL, 2005: pp. 805–850. https://doi.org/10.1201/9781420027686.
152. T. Serra, X. Casamitjana, Modelling the aggregation and break-up of fractal aggregates in a shear flow, *Appl. Sci. Res.* 59 (1997) 255–268. https://doi.org/10.1023/A:1001143707607.
153. G. McConnachie, Turbulence intensity of mixing in relation to flocculation, *J. Environ. Eng.* 117 (1991) 731–750. https://doi.org/10.1061/(ASCE)0733-9372(1991)117:6(731).
154. K. Haralampides, A.J. McCorquodale, B.G. Krishnappan, Deposition properties of fine sediment, *J. Hydraul. Eng.* 129 (2003) 230–234. https://doi.org/10.1061/(ASCE)0733-9429(2003)129:3(230).
155. A.K.C. Yeung, R. Pelton, Micromechanics: A new approach to studying the strength and breakup of flocs, *J. Colloid Interface Sci.* 184 (1996) 579–585. https://doi.org/10.1006/jcis.1996.0654.
156. B. Dobias, W. Von Rybinski, Stability of dispersions, in: B. Dobias, X. Qiu, W. Von Rybinski (eds.), *Solid-Liquid Dispersions*, Marcel Dekker, New York, NY, 1999: pp. 244–278. https://doi.org/10.1201/9781482269727.
157. S.J. Peng, R.A. Williams, Control and optimisation of mineral flocculation and transport processes using on-line particle size analysis, *Miner. Eng.* 6 (1993) 133–153. https://doi.org/10.1016/0892-6875(93)901[illegible]0-A.
158. L.E. Neumann, T. Howes, Aggregation and breakage rates in the flocculation of estuarine cohesive sediments, in: J.P.Y. Maa, L.P. Sanford, D.H. Schoellhamer (eds.), *Estuarine and Coastal Fine Sediment Dynamics*, Elsevier, Amsterdam, 2007: pp. 35–54.
159. L. Oshinowo, E. Elsaadawy, R. Vilagines, CFD modeling of oil-water separation efficiency in three-phase separators, in: J.E. Olsen, S.T. Johansen (eds.), *Selected papers from 10th International Conference on Computational Fluid Dynamics in the Oil & Gas, Metallurgical and Process Industries*, SINTEF Academic Press, 2015.
160. I. Nopens, *Improved Prediction of Effluent Suspended Solids in Clarifiers through Integration of a Population Balance Model*, Taylor & Francis, Toulouse, France, 2007.
161. A.R. Heath, P.T.L. Koh, Combined population balance and CFD modelling of particle aggregation by polymeric flocculant, in: *Third International Conference on CFD in the Minerals and Process Industries CSIRO*, Melbourne, Australia, 2003.
162. E. Torfs, *Different Settling Regimes in Secondary Settling Tanks: Experimental Process Analysis, Model Development and Calibration*, PhD, Ghent University, 2015.
163. E. Torfs, M. Vesvikar, I. Nopens, Improved predictions of effluent suspended solids in wastewater treatment plants by integration of a PBM with computational fluid dynamics, in: *5th Population Balance Modelling Conference*, Bangalore, India, 2013.
164. B.J. Lee, F. Molz, Numerical simulation of turbulence-induced flocculation and sedimentation in a flocculant-aided sediment retention pond, *Environ. Eng. Res.* 19 (2014) 165–174. https://doi.org/10.4491/eer.2014.19.2.165.

2 The Role of Fluid Dynamics and Fluid–Particle Interaction on Fouling in Membrane-Based Separation Systems

Suzanne Yala

2.1 INTRODUCTION

Membrane-based separation processes have increasingly been applied across various fields. This is due to their advantages such as excellent permeate quality, smaller footprint, and insignificant chemical composition [1]. Despite the highlighted benefits, membrane-based processes still face a major challenge of membrane fouling. Membrane fouling, which is defined as the deposition of material onto the surface and within the pores of membrane, reduces the process efficiency by increasing the resistance to flow and consequently decreasing the permeate flux or increasing the transmembrane pressure (TMP). In addition, membrane fouling results in an increase in the overall operational cost due to the frequent membrane cleaning required during the separation process.

In order to maintain the productivity of membrane processes, various ways of controlling membrane fouling have been investigated and are being applied. They include feed pretreatment, membrane surface modification, and optimisation of fluid dynamic conditions. It should be noted that the mitigation methods of membrane fouling are applied depending on the parameters that exacerbate fouling in particular membrane systems. Therefore, understanding the major factors affecting membrane fouling is key to choosing the appropriate fouling control strategies.

This chapter focuses on one of the key parameters that has an impact on fouling in membrane-based processes, that is, fluid dynamics and its interaction with foulants. First, the important aspects of membrane fouling are outlined, then an overview of fluid dynamics and fluid–particle interactions in a

DOI: 10.1201/9781003518426-2

membrane system is briefly presented. Lastly, a detailed discussion and review of previous studies on the impact of fluid–particle interactions are presented.

2.2 FUNDAMENTALS OF MEMBRANE FOULING

2.2.1 Fouling Phenomena

Membrane fouling is defined as the adsorption and deposition of colloidal and soluble matter on a membrane surface and within membrane pores [2]. Depending on the filtration mode, membrane fouling results in a decline in permeate flux (in constant pressure mode) or an increase in TMP due to hydraulic resistance created by the fouling layer [3]. However, most commercial membrane systems are operated in constant flux mode; hence, the illustration of the fouling phenomenon will be focused on membrane systems operated under constant flux mode.

Fouling layer development, especially when it involves compressible foulant, usually occurs in three main phases as illustrated in Figure 2.1 [4–6]:

i. First, there is a 'conditioning fouling stage' where an initial rapid increase in TMP occurs due to adsorption of foulant within the membrane pores.
ii. The conditioning stage is followed by a 'slow and steady fouling process' which is characterised by a slow increase in TMP due to partial or complete blockage of membrane pores and also slight adsorption of foulants on the membrane surface.
iii. The final stage is referred to as 'TMP jump'. In this phase, there is a sudden increase in TMP due to the accumulation of foulants on the membrane surface and also on other deposited foulants. The jump in TMP is also attributed to the compression of deposited foulants due to permeation drag, as will be discussed in Section 2.4.

The deposition of foulants on a membrane and the subsequent increase in TMP are highly dependent on fluid dynamics [7, 8]. Fouling occurs when the convective fluid flow of foulants towards the membrane

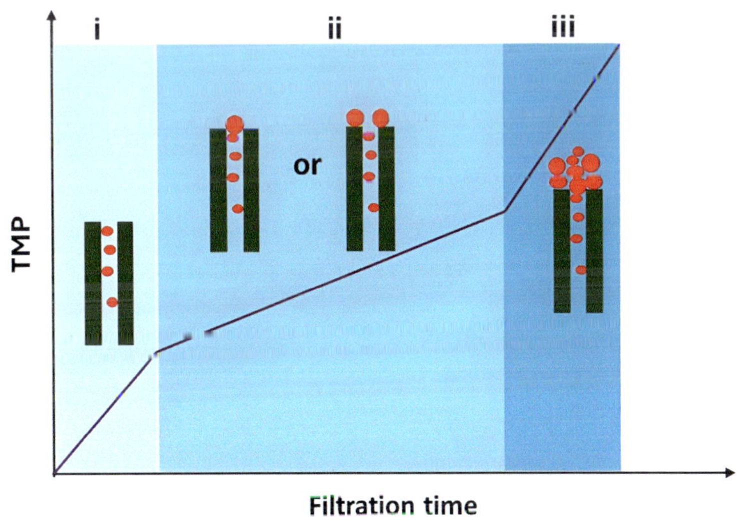

FIGURE 2.1 The change in transmembrane pressure with time during a membrane-based separation process, illustrating the stages of fouling layer development.

is higher than the back transport of the foulants from the membrane surface to the bulk feed [9]. This is influenced by both the flow of fluid and its interaction with foulants. Hence, fluid dynamics and fluid–foulant interaction play a key role in membrane fouling. More details on this are outlined in Section 2.3.

2.2.2 Classification of Membrane Fouling

Understanding the types of membrane fouling is essential in identifying how the fouling process occurs, the factors that exacerbate membrane fouling, and consequently, the ways of minimising it [10, 11]. Membrane fouling can be classified based on various criteria, as illustrated by Figure 2.2. Most membrane researchers categorise membrane fouling based on the strength of attachment of the foulant onto the membrane and the characteristics of the foulant, while others consider the mechanism with which the membrane is fouled [4, 12, 13].

According to the attachment strength of foulants on a membrane, membrane fouling can be divided into reversible, irreversible, or irrecoverable [14]. Reversible fouling, which is known as removable, is a fouling layer that is loosely attached to a membrane and can be removed by physical cleaning methods such as air scouring, backwashing, and water flushing. In some cases, the membrane fouling can only be removed by chemical cleaning because of its strong attachment to the membrane [15]. This type of fouling is referred to as irreversible fouling. It mostly occurs during a long continuous membrane separation process where a reversible fouling layer gets consolidated and turns into irreversible fouling. An irrecoverable fouling layer corresponds to a layer that can neither be removed by physical cleaning nor chemical cleaning [16]. Hence, it is sometimes referred to as irremovable fouling.

Membrane fouling categorised based on the properties of foulants includes inorganic fouling, organic fouling, and biofouling [17]. Inorganic fouling is caused by precipitation and crystallisation of inorganic compounds (such as salts and metal oxides) on the membrane. A fouling layer as a result of the deposition of organics such as polysaccharides, humic acid, and protein is referred to as an organic fouling layer. Biofouling is the accumulation and growth of bacteria on a membrane over a period of time.

Membrane fouling can be classified as pore constriction, intermediate pore blocking, complete pore blocking, or cake layer formation, and this depends on the mechanism of fouling as shown in Figure 2.3 [1, 18]. This classification of membrane fouling is dependent on the size of the foulant in relation to the membrane pore size. If the size of the foulant is smaller than the pore size of the membrane, the foulant will get adsorbed and deposited within the membrane pores [19], thus constricting the pores. This type of fouling is referred to as pore constriction, pore narrowing, or standard pore blocking. It mostly occurs at the initial stage of a membrane separation process. A membrane feed can also contain foulants that have

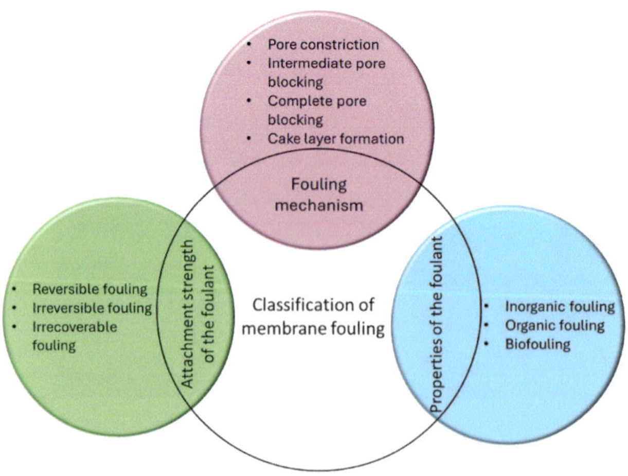

FIGURE 2.2 Classifications of membrane fouling.

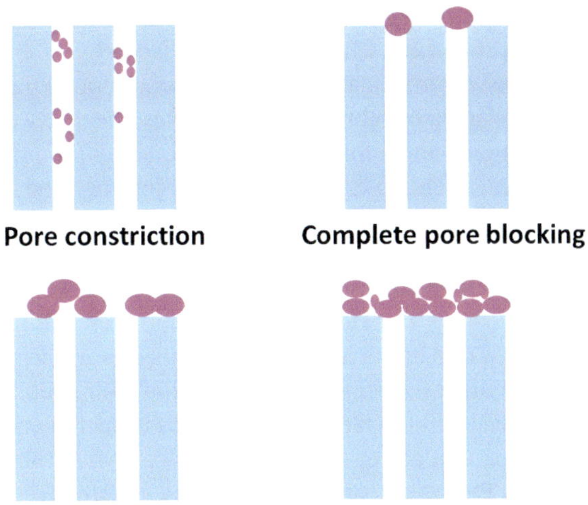

FIGURE 2.3 Types of membrane fouling based on fouling mechanisms.

TABLE 2.1 The Relationship between the Different Classification of Membrane Fouling

PROPERTIES OF FOULANT	*FOULING MECHANISM*	*ATTACHMENT STRENGTH OF FOULANT*
Inorganic fouling	Pore constriction	Irreversible/irrecoverable fouling
Organic fouling	Cake layer formation	Reversible fouling
	Pore constriction	Irreversible/irrecoverable fouling
Biofouling	Cake layer formation	Reversible fouling
	Pore constriction	Irreversible/irrecoverable fouling

a size that is equal to or close to the membrane pore size that will either completely block or partially block the membrane pores without lying on other foulants during the separation process. These types of fouling are known as complete pore blocking and intermedial/partial blocking, respectively. Foulant of relatively large size compared to the membrane pores will form a cake layer. In this type of fouling, the foulant deposits on the membrane surface and on other foulants. It always occurs in the later stages of a membrane separation process when all the membrane pores have been blocked.

Apart from being in individual categories, the different types of membrane fouling are related to each other as illustrated in Table 2.1 [4].

It should be noted that fouling layers classified based on the strength of attachment of foulant onto the membrane and fouling mechanism are influenced by fluid dynamics and fluid–particle interactions. For example, several authors have reported that an irreversible fouling layer is usually observed at high applied flux/pressure due to high fluid permeation drag [20, 21]. On the other hand, reversible fouling mostly occurs at relatively low applied fluxes. Therefore, understanding the influence of fluid dynamics and fluid–particle interactions on membrane fouling, which is the focus of this chapter, is critical in identifying appropriate mechanisms of mitigating the specific types of fouling.

2.2.3 Factors Affecting Membrane Fouling

Understanding the major factors affecting the deposition of foulant can help in mitigating and minimising membrane fouling. Generally, these factors can be categorised into three groups: membrane surface characteristics; feed and foulant properties; and fluid dynamic conditions [1, 22]. These parameters affect the fouling process individually or as combination of two or three factors.

One of the factors that affects the interaction between foulant and a membrane is the surface properties of the membrane. The properties, which include surface charge (measured in terms of zeta or streaming potential), membrane roughness, and the electron donor potential, which have an effect on the energy barrier between foulants and a membrane. For example, a decrease in zeta potential decreases the energy barrier between a membrane and a foulant and consequently promotes foulant deposition. Hence, membranes with low zeta potential tend to experience severe fouling. The increase in membrane surface roughness also reduces the height of the energy barrier, thus increasing the attractive energy of the foulant to the membrane. These few examples of the impact of membrane characteristics on fouling highlight the key role these parameters play in membrane fouling. However, it should be noted that membrane properties do not affect fouling individually. Its effect is in relation to feed and foulant properties. In addition, membrane properties only play a role in the initial stages of membrane separation processes. Under long separation processes, foulant–foulant interactions are dominant compared to membrane–foulant interactions. Therefore, the focus always shifts to feed characteristics in such scenarios.

Foulant and feed properties that affect membrane fouling include floc size, feed pH, and surface charge. Studies have shown that foulants of smaller size tend to adhere to the membrane compared to their larger counterpart, especially on a rough membrane surface. This is because foulants with smaller floc size have a higher interactive force and can easily be carried by fluid flow to the membrane surface during filtration. Analysis of physicochemical interactions by various researchers has also shown that the adhesion of foulants onto the membrane surface is high in acidic conditions compared to alkaline conditions. They attributed this to a decrease in energy barrier at low pH values.

Apart from the membrane, feed, and foulant properties, fluid dynamic conditions also play a major role in membrane fouling. The main factors that control fluid dynamics in membrane-based systems are cross-flow velocity (or aeration) and permeation drag. Cross-flow and aeration promote back transport of foulants from the membrane, which results in less foulant deposition. However, a high rate of cross-flow or aeration can also result in severe fouling as it may break down foulants into small particle sizes. This in turn increases the fouling rate. Permeation drag, which is generated by applied pressure/flux, transports foulants towards the membrane surface. An increase in applied pressure/flux would result in an increase in the amount of foulant deposited on a membrane. This is due to the increase in the permeation drag. Permeation drag also affects other aspects of membrane fouling, such as compression of fouling layer, fouling layer characteristics, and reversibility of fouling layers. This brief description shows that fluid dynamics play a major role in both promoting and limiting membrane fouling. Detailed discussions on this are presented in Sections 2.3 and 2.4.

2.2.4 Strategies for Membrane Fouling Control

In order to have optimal performance in membrane-based separation processes, membrane fouling needs to be controlled. There are two groups of techniques that have been widely researched and employed to control fouling, that is, minimising techniques and fouling removal techniques [23]. Based on the factors affecting fouling outlined in Section 2.2.3, techniques of minimising fouling can be categorised into three major groups—membrane modification, feed pretreatment, and optimisation of fluid dynamics conditions [24–26]. Examples of these strategies and the mechanisms employed by each of them are briefly described in Table 2.2.

The techniques of membrane modification, feed pretreatment, and optimisation of fluid dynamic conditions help in limiting fouling before it occurs [41]. Membrane modification (through incorporation of nanoparticles or blending with zwitterions) and feed pretreatment (coagulation, adsorption, or oxidation) reduce the physicochemical interactions between a membrane and the feed and thus limit the deposition of foulants on the membrane [12, 42]. The optimisation of fluid dynamics conditions promotes back transport of foulants from the membrane to the bulk feed, while limiting convection flow of foulant to the membrane [17, 43], hence delaying the deposition of foulants on the membrane.

TABLE 2.2 Strategies of Minimising Membrane Fouling

MINIMISING FOULING	*EXAMPLES*	*MECHANISM*	*REFERENCES*
Membrane modification	Incorporation of nanoparticles such as titanium dioxide and silver	Increases electron donor tension potential and zeta potential of membrane	[27, 28]
	Coating/blending/grafting with organic modifiers such as zwitterions	Increases the energy barrier between foulant and membrane. Makes the membrane hydrophilic, thus creating a hydration barrier to resist foulant adhesion	[29, 30]
Feed pretreatment	Adsorption using activated carbon	Reduces the attractive interactions of foulant by adsorbing organics in the feed and increasing zeta potential	[31, 32]
	Adding coagulants such as ferric chloride to feed	Decreases fouling by enlarging foulant size, decreasing organics	[33, 34]
	Oxidation	Fouling is minimised by decreasing the molecular weight of foulant, thus reducing attractive interactions	[35, 36]
Optimising fluid dynamic conditions	Optimising cross-flow and aeration conditions	Facilitates back transport of foulant and thus reduces deposition on the membrane	[37, 38]
	Sustainable flux	By operating below a defined critical flux/pressure. Hence, minimising the deposition of foulant	[39, 40]

TABLE 2.3 Membrane Cleaning Methods

REMOVAL OF FOULING LAYER	*EXAMPLE*	*MECHANISM*	*REFERENCES*
Physical cleaning	Backwashing	It involves reversing the flow of filtration, thus dislodging foulants	[48, 50]
	Membrane relaxation	By temporary pausing filtration, the reversible foulants move back to the bulk feed	[51, 52]
Chemical cleaning	Soaking in acid, bases, or disinfectant	Reacts with foulant and enhances their solubility, thus breaking their bond with the membrane	[38, 53]
Physiochemical cleaning	Chemically enhanced backwash	Reverse flow of water containing chemicals, thus solubilising and detaching fouling layers	[54]
Biological cleaning	Quorum quenching	Limits communication between microbes, thus stopping the production of auto-inducible factors that cause biofouling	[49, 55]
	Bacteriophages	Infects and destroys bacteria, thus inhibiting the formation of biofouling layer	[56, 57]

Regardless of sufficient feed pretreatment, proper membrane modification, and optimisation of operating conditions, fouling is inevitable with time; therefore, membrane cleaning is required [13, 44]. Membrane cleaning methods can be grouped into four main types (Table 2.3): physical methods, chemical methods, physiochemical methods, and biological methods [44, 45]. Physical cleaning methods (air scouring, backwash, and membrane relaxation) employ mechanical or hydraulic forces to remove foulants from the membrane surfaces during membrane filtration operations [1]. However, it is only able to remove reversible fouling layers. Irreversible foulants are usually removed by chemical cleaning [46]. It involves the use of chemicals such as acids, bases, and disinfectants to solubilise and detach foulants from the membrane. Despite its effectiveness, frequent chemical cleaning damages the membrane and hence reduces its life span. To enhance the effectiveness of physical cleaning methods in foulant removal, some studies have combined the physical method with the use of chemicals [47, 48]. This cleaning method is referred to as physiochemical cleaning method. One of the major examples is the chemically enhanced backwash. Another cleaning method that has recently gained attention is the biological cleaning method [9, 49]. This method, which includes quorum quenching and bacteriophages, is used to specifically mitigate biofouling by destroying bacteria or limiting the communication between them.

2.3 FLUID DYNAMICS AND FLUID–PARTICLE INTERACTIONS IN MEMBRANE-BASED SEPARATION PROCESSES

2.3.1 Overview

In membrane systems, a driving force (such as applied flux) is usually applied to induce separation. The driving force always results in a fluid flow that imposes permeation drag force on foulants towards the membrane, where they are retained [58]. This causes the area near the membrane surface to have a high concentration of particles compared to the bulk feed, thus inducing a back flow of foulants through Brownian motion and the inertia lift mechanism [59]. In cross-flow systems, the back transport can also occur through a shear-induced mechanism. In this chapter, the interplay between convective permeation drag and back transport of foulant is what is referred to as fluid dynamics in membrane-based separation systems.

During the initial few seconds of the membrane separation process, a condition is usually reached where the convective permeation drag equals to back transport [60]. At these conditions, a concentration boundary layer, which affects fluid flow, forms near the membrane surface as illustrated in Figure 2.4. This phenomenon is commonly known as concentration polarisation and is a precursor of membrane fouling [61, 62].

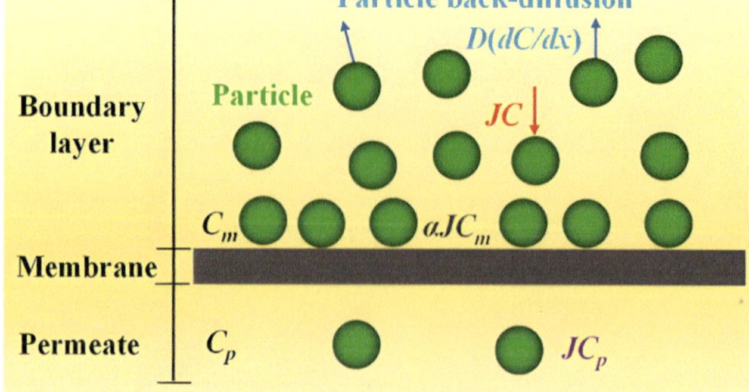

FIGURE 2.4 An illustration of fluid transport and concentration polarisation phenomena in a membrane-based separation process [60].

As the membrane separation process proceeds, the convection flow of foulants becomes greater than the back transport, thus resulting in the accumulation of foulants on the membrane surface and pores [63, 64], thus forming a membrane fouling layer. The formation of a membrane fouling layer is always pronounced at a longer separation period and at high applied fluxes. More on this and its effects will be discussed in the subsequent section.

Apart from fluid dynamics, foulant–membrane interactions and foulant–foulant interactions also play a key role in the occurrence of membrane fouling. These interactions result in energy barrier forces that attempt to prevent the deposition of foulant [65, 66]. Hence, the accumulation of foulants on membranes during the separation process is also influenced by the interaction between permeation drag and the energy barrier forces.

Studies have used different approaches to represent these interactions between drag and barrier forces [67–69]. Tang and co-workers, for instance, used the balance between the permeation drag force that advances foulant deposition and the barrier force that limits fouling [64, 66]. However, this approach relied on the assumption of a constant interaction barrier force in respect of the foulant–membrane distance. However, this could not hold, as the value of the barrier forces continued to change as filtration progressed. Recently, another research group has investigated the interplay between these forces, using a collision-attachment model adapted from the field of coagulation [65, 67]. In this model, the membrane is viewed as a large, boundless matter, and fouling as a string of collisions of foulants with the membrane. As illustrated in Figure 2.5, foulants can get convected towards, and bump up against, the membrane under the effect of drag force. The amount of foulant per unit time that is successfully attached to the membrane is controlled by the collision frequency and attachment efficiency, and is represented mathematically in Equations (2.1) and (2.2) [70]:

$$dm_f / dt = \gamma J C_d \tag{2.1}$$

where m_f and t are the amount of foulant deposited and the time of filtration, respectively, J represents the flux, and C_d is the concentration of the suspension being filtered. The term JC_d represents the frequency with which foulants bump up against the membrane, while γ is the likelihood of successful attachment of the foulant for every collision. γ is calculated from the Boltzmann distribution law, which includes the combined role of permeation drag, barrier forces, and concentration polarisation on the deposition of foulants, and is represented as [67]:

$$\gamma = \frac{1}{1 + \exp\left(\Delta E_b / k_B T - \Delta E_d / k_B T - J / k_m\right)} \tag{2.2}$$

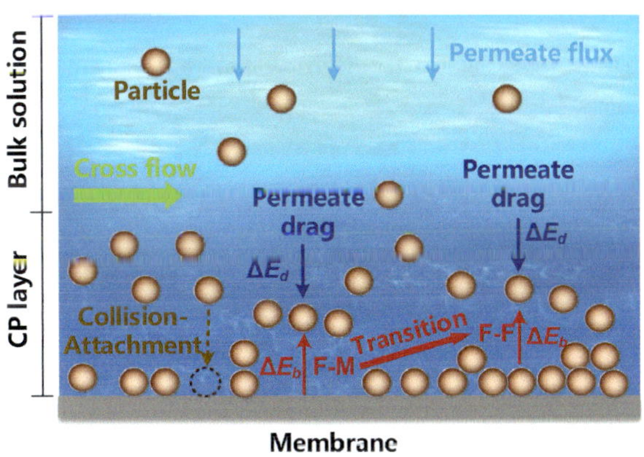

FIGURE 2.5 Collision and deposition of foulants on a membrane as a result of the interaction between permeation drag and energy barrier forces. Adapted with permission from Elsevier B.V. [68].

where k_B represents Boltzmann's constant, T is the solution temperature, ΔE_b is the energy barrier force associated with foulant–membrane and foulant–foulant interactions, ΔE_d is the permeation drag force, J haracterises the permeation flux, and k_m is the mass transfer coefficient.

In describing fluid–particle dynamics and membrane fouling, the interaction between the permeation drag and the energy barrier force is always related to the critical flux [66, 68, 71]. This relation identifies three scenarios [71]:

- At a flux below the critical value, the drag is fairly low compared to the energy barrier force. Hence, foulant deposition is negligible. This case mostly results in concentration polarisation.
- At the critical flux, where the permeation drag is equal to the barrier force, the deposition of particles increases. Two types of layers occur, namely a layer of uniformly and loosely packed foulants at the membrane surface, and a concentration polarisation layer.
- When the permeation drag is greater than the barrier forces (i.e., above the critical flux), severe fouling occurs. The distance between the foulants decreases, resulting in a packed layer. As the foulant–foulant interactions increase, the deposits are compressed, thus forming an irreversible layer.

2.3.2 Impact of Fluid Dynamics and Fluid–Particle Interactions on Membrane Fouling

In their search for strategies to minimise membrane fouling in filtration systems, researchers have investigated the various effects of fluid–particle dynamics due to permeation drag on fouling [72, 73]. These effects are illustrated in Figure 2.6.

2.3.2.1 Mass Transfer and Foulant Deposition

Studies have also demonstrated that high initial fluxes can cause more severe fouling than when operating at low fluxes. This has been attributed to a relatively stronger permeation drag that exceeds back transport and the energy barrier force, thereby increasing the chances of successful attachment during

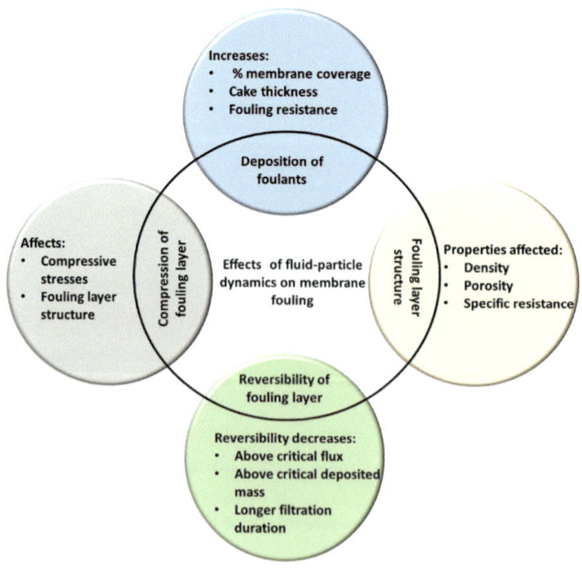

FIGURE 2.6 Effects of fluid dynamics and fluid–particle interactions on membrane fouling [12].

collision [74, 75]. And, in turn, this increases the amount of foulant deposited [67, 76]. For instance, Liu et al., in analysing the relationship of fouling to permeation flux and energy barrier, observed that the amount of foulant deposited on the membrane (dm_f / dt) increased from 10^{-9} to 10^{-5} g/m²s, when the initial flux was increased from 10 to 70 µm/s [68]. They attributed the increase in foulant deposition to the increase in the likelihood of successful attachment (Equation 2.1), which is related to the value of permeation drag relative to energy barrier (Equation 2.2). In another study, where the effect of permeation drag was evaluated, it was reported that a membrane in an operation mode where flux applied had a more fouling propensity compared to a membrane system where there was no flux [76]. They linked the high membrane fouling propensity to the foulant being able to overcome the energy barrier forces due to permeation drag acting on them. Zhao and his colleagues, in their evaluation of the role of initial flux relative to energy barrier, observed stable and mild flux decline in low initial flux runs (30 L/m²h) compared to runs done at high initial flux (300 L/m²h) where severe fouling occurred [74]. They linked the severity of fouling at high initial flux to pronounced permeation drag, increased collision frequency (JC_d in Equation 2.1), and high attachment probability (which is plotted for the different runs). Hence, it can be concluded that the relationship between fluid dynamics (in terms of permeation drag) and fluid–particle interactions (represented by energy barrier) impacts the amount of foulant that is deposited on the membrane.

The effect of fluid dynamics and fluid–particle interactions is always quantified in different ways by various researchers, as shown in Table 2.4. The quantification parameters include the percentage or fraction of membrane coverage or the mass of foulant deposited, cake thickness, and fouling resistance.

(a) Percentage of membrane coverage/mass of foulant deposited

Various studies have quantified the effect of fluid–particle interactions on the mass transfer of foulants onto a membrane as a fraction of the membrane covered by foulants, the available membrane area ratio, or the mass of foulant deposited on the membrane. For example, in describing the flux variation due to an increase in applied pressure and, consequently, in fluid–particle interactions, Hou and co-workers [85] quantified the foulant deposited on polyethersulfone (PES) microfiltration and ultrafiltration membranes during the filtration of bovine serum albumin (BSA) as an available membrane area ratio. From Glen Bolton's model equation, the researchers developed the expression shown below in Equation (2.3), which

TABLE 2.4 Quantification of the Effect of Fluid Dynamics and Fluid–Particle Interactions on Mass Transfer and Deposition of Foulants

QUANTIFICATION PARAMETER	*ADVANTAGES*	*LIMITATIONS*	*REFERENCES*
Percentage of membrane coverage/available membrane area ratio	No specialised equipment is required, hence can easily be used	Cannot be used in cases where there is more than one fouling layer formed	[77–79]
Cake thickness	Can be used in cases where there are multiple fouling layers deposited	Require specialised equipment like a fluid dynamic gauge to quantify Not applicable in the membrane filtration of compressible foulants	[80–82]
Fouling resistance	Can be used in a wide range of cases, such as in the filtration of compressible foulants and in the deposition of multiple fouling layers Does not require a specialised equipment	n/a	[79, 83, 84]

relates the available membrane area ratio (A/A_0) to the permeation flux (J) for a filtration process performed in constant pressure mode [85]:

$$J/J_o = R_o A / R A_o \tag{2.3}$$

where J_O represents the initial flux ($m^3/(m^2h)$), R characterises the sum of the membrane resistance and fouling resistance (m^{-1}), and R_o is the initial resistance (m^{-1}). A and A_0 are the available membrane area (m^2) and the initial membrane area (m^2), respectively. The equation was used to compute the membrane area ratio available at different levels of applied pressure. The researchers observed that the A/A_0 decreased more quickly at high applied pressure. This they attributed to the high driving force, which resulted in the quick accumulation of foulants on the membrane surface.

In a similar study on the variation in transmembrane pressure (TMP) in a microfiltration membrane system operated in constant flux mode, Du and colleagues [79] quantified the influence of permeation drag on foulant deposition by measuring the membrane area ratio (A/A_0) available over time for different applied fluxes. Six permeation fluxes, namely 60, 70, 80, 90, 100, and 120 L/m^2h (LMH), were investigated. The ratio A/A_0 significantly decreased as permeation flux increased. The ratio decreased from 0.956 to 0.802 when the 60 LMH flux was doubled. This observation was linked to the high rates of particle transfer from the solution to the membrane surface.

In another investigation by Di and colleagues [86], a 20 mg/L of latex suspension was filtered through a 10 kDa PES membrane, and the impact of permeation flux on membrane percentage coverage was evaluated. Comparing the change in flux to surface coverage, the researchers found that coverage increased more than twofold (from 10% to 24%) when the flux was increased from 12.5 LMH to 25 LMH. A threefold increase in the flux (from 25 to 75 LMH) resulted in only a 1.5-fold increase in coverage. This finding indicated the limit of quantifying the effect of permeation drag by means of membrane surface coverage.

(b) Cake thickness

Alternatively, some researchers have used cake thickness to quantify the effects of fluid–particle dynamics on fouling layer deposition [87]. An example of this is the experimental investigation by Arandia and colleagues [80], who monitored the fouling of a PES microfiltration membrane under different operating conditions. Using a fluid dynamic gauge (FDG), the researchers measured the layer thickness of microcrystalline cellulose during the filtration process at different levels of applied pressure, with the cross-flow velocity and feed concentration remaining constant. A plot of the cake layer thickness against applied pressure showed that an increase in pressure from 200 to 400 mbar resulted in an increase in the cake layer thickness from 394 to 628 μm.

Mattsson's group [88], too, studied the impact of applied TMP on the accumulation of foulants, also using FDG to measure the cake layer thickness during the microfiltration of kraft lignin. They found that the measured thickness of the deposits increased with the applied pressure. The layer thickness increased from 45 to 85 μm when the filtration TMP increased from 50 to 200 mbar. They attributed this increase in thickness to a high permeation flux as a result of high applied pressure, which, in turn, resulted in a larger amount of foulant being transferred to the membrane.

However, for compressible foulants (such as bacteria and yeast), an increase in permeation drag does not necessarily translate to an increase in cake thickness [89], and the mass deposited on the membrane may not be directly related to the fouling layer thickness. Dreszer and colleagues demonstrated this claim [90]. In their work, they observed that a biofilm formed at a flux of 20 LMH was thicker than that developed under a permeation flux of 100 LMH. This implied that the fouling layer developed at 100 LMH was compressed. Jiang's research group [82] made similar observations during the separation of algae by means of a membrane filtration system. They observed a cake thickness of 26 μm at an operating TMP of 10 kPa, but a thinner cake of 18 μm at a higher operating pressure of 18 kPa. Judging from these conflicting findings, fouling layer thickness alone may not be an accurate measure of the impact of fluid dynamics on membrane fouling.

(c) Fouling resistance

Other researchers have attempted to use fouling resistance as a measure to determine the impact of permeation drag relative to the energy barrier on foulant deposition. A higher permeation drag is reported to increase the fouling resistance because of an increase in the amount of matter deposited on the membrane [91, 92]. As shown in Equation (2.4), the fouling resistance and the amount of foulant deposited can be expressed as a correlation [93]:

$$R_f = \alpha cV / A = \alpha\omega \tag{2.4}$$

where R_f represents the fouling resistance (m^{-1}), α is the specific cake resistance (m/kg), and ω is the amount of foulant deposited per unit area of membrane surface (kg/m^2). Thus, the change in fouling resistance as a result of deposition is derived from Equation (2.4) and expressed as [94]:

$$dR_f / dt = \alpha d\omega / dt \tag{2.5}$$

The influence of permeation drag on the deposition of foulants, and therefore on the fouling resistance in membrane systems, is always described by Darcy's law, which is represented as [95, 96]:

$$J = TMP / \mu\left(R_m + R_f\right) = TMP / \mu\left(R_m + \alpha\omega\right) \tag{2.6}$$

where J is the flux (m/s or L/m^2h), TMP is the transmembrane pressure (Pa), μ is the fluid viscosity (Pa.s), and R_m and R_f represent the membrane resistance (m^{-1}) and the fouling resistance (m^{-1}), respectively.

Following the same approach, Baptista and co-workers [84] used Darcy's correlation to describe the relationship between resistance and cake buildup during ultrafiltration of *Eucalyptus* bark extract. According to their analysis, the fouling resistance doubled, and in some instances even tripled, when TMP increased from 3 to 8 bar. Jiraratananon and Chanachai [83] reported a similar trend of increasing fouling resistance with increasing TMP in their study of the performance of alumina hollow-fibre membrane in the purification of genipap fruit extract. In another related study, Du and colleagues [97] reported an increase in total resistance from 0.31×10^{12} m^{-1} to 0.63×10^{12} m^{-1} when the operating flux was increased from 50 to 200 LMH during the ultrafiltration of TiO_2 particles combined with humic acid. All the aforementioned researchers attributed the positive correlation between fouling resistance and applied pressure or flux to the increase in the transportation of foulants towards the membrane surface.

2.3.2.2 Compression of Fouling Layer

Studies have also shown that, apart from increasing the amount of foulant deposited, greater applied pressure or flux can also compact and subsequently increase the specific cake resistance (α) of fouling layers on a membrane [93, 98–100]. This, in turn, results in a high overall fouling resistance (R_f). In an investigation by Costa and co-workers [99], a CA1 membrane was operated at a pressure 25 times greater than another membrane (CA5). The CA1 membrane exhibited a high specific fouling resistance value of 1.6×10^{11} m/mg against the CA5 membrane's 3.7×10^9 m/mg, yet had a smaller mass deposit of 13 mg versus 22 mg. The high specific cake resistance exhibited by CA1 despite the lower mass deposited was linked to the compaction of the fouling layer because of the high applied pressure. Robles and colleagues [101] also reported a considerable increase in the specific cake resistance from approximately 2×10^{13} to 25×10^{13} m/kg when the TMP was raised from 0.1 to 0.3 bar. Dreszer's research group [102] observed a similar trend for a biofilm formed on a microfiltration membrane. The resistance of the biofilm was reported to have increased from approximately 40×10^{11} m^{-1} to 125×10^{11} m^{-1} when the imposed flux was increased from 20 to 60 LMH. This was accompanied by a 40% decrease in biofilm thickness, indicating that the biofilm had become more compact with increasing permeation flux. From these investigations, one can

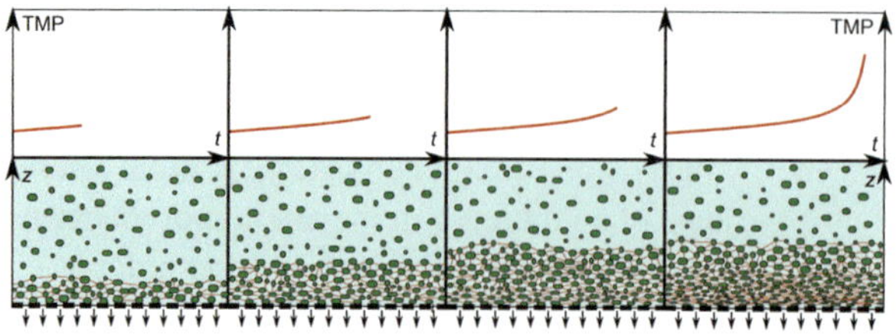

FIGURE 2.7 An illustration of the effect of permeation drag relative to energy barrier on the compression of membrane fouling layers. Reused with permission from John Wiley & Sons [103].

conclude that fluid dynamics relative to energy barrier forces in membrane systems not only affects foulant deposition but also causes compression and consolidation of the fouling layer.

The effect of fluid–particle dynamics due to permeation drag on the compaction of the fouling layer is well illustrated in Figure 2.7 by Skinner and colleagues [103], who found that at the start of a membrane filtration process, the TMP increased gradually due to the accumulation of foulants. As the applied pressure increased, the deposited foulants consolidated into a dense structure because of an increase in foulant–foulant interactions. To maintain a constant flux, the permeation drag was also increased, resulting in further compression. Eventually, the compression of the cake layer dominated the formation of fouling. Hence, a steep increase in TMP was observed.

2.3.2.3 Fouling Layer Characteristics

Compression caused by applied permeation drag forces in membrane filtration is reported to have a major influence on the structure and characteristics of the fouling layer. Studies have shown that permeation drag through a fouling layer leads to changes in the layer's thickness, porosity, and hydraulic resistance. This, in turn, adversely affects the filtration and cleaning performance of membrane processes. Hence, this section reviews previous investigations on the effects of fluid–particle dynamics on the various structural characteristics of the fouling layer, and how these affect the performance of the membranes.

(a) Thickness

Previous studies have revealed that an increase in permeation flux or applied pressure may result in a reduction in the thickness of a compressible fouling layer. This change in thickness is related to the compaction of the cake layer, i.e., the previously formed biomass layer occupies a smaller volume than it initially did. Using optical coherence tomography (OCT), Dreszer and colleagues [102] observed the compaction and consequent decrease in thickness of a biofilm formed on an ultrafiltration membrane under stepwise flux changes. At triple the flux, the biofilm layer thickness decreased by half. Another investigation evaluated the effect of two imposed fluxes, one above and the other below the critical flux of approximately 15 LMH, on the thickness and density of a dynamic fouling layer on a nonwoven membrane bioreactor [104]. At fluxes below 15 LMH, a thick but loose biofouling layer developed on the membrane surface. However, when the imposed flux exceeded the critical level, a thin but dense fouling layer formed. This significantly reduced the permeability of the membrane, and the high compactness of the fouling layer caused severe membrane fouling. A study by Valladares Linares and colleagues [105] on the compaction and relaxation of biofilms in membrane systems demonstrated that an increase in flux from 20 to 60 LMH led to a 49% reduction in the fouling layer thickness. When they reverted to a flux of 20 LMH, the initial thickness of the biofilm was not restored. The biofilm thickness decreased by 21%, while the resistance to flow increased by 16%. This was attributed to a decrease in void spaces within the biofilm layer as the applied flux increased. The researchers concluded that an increase in permeation drag could result in compaction of the biofilm layer, a resultant decrease in biofilm thickness, and, consequently, a decline in membrane performance. In summary, therefore, an increase in the permeation drag (i.e., permeation flux or applied pressure) on a fouling layer composed of compressible particles results in a decrease in the interparticle distance because of an increase in foulant–foulant interactions. This, in turn, decreases the cake layer thickness.

This compression of the fouling layer under high permeation drag also explains why a thinner layer can have a greater resistance to flow than a thicker layer. This is inconsistent with the effect of fluid–particle dynamics on the deposition of particles, namely that a high flux results in thicker layer deposition, and vice versa. Hence, as mentioned in Section 2.3.2.1, the fouling layer thickness may not be an accurate parameter for characterising the effect of fluid–particle dynamics on membrane fouling [102].

(b) Porosity

Another structural fouling layer property that is greatly affected by permeation drag is porosity. Many researchers have observed that when a fouling layer in membrane filtration systems is subjected to a high permeation drag force (such as high applied TMP or flux), it leads to a reduction in the layer's porosity. This is attributed to an increase in foulant–foulant interactions at high fluxes (or pressure), which results in a reduction in the inter-foulant distance [64]. In their evaluation of the characteristics of membrane biofouling layers formed at different fluxes during drinking water treatment, Sun and co-workers [106] reported that the porosity of the layers decreased as the flux increased from 50 to 70 LMH. Similar results were obtained by Abbasi and co-workers [107]. During their modelling of membrane fouling and flux decline in the filtration of oily wastewater through a mullite ceramic membrane, they observed that the fouling layer porosity decreased from 25.68% to 14.98% when the operating pressure increased from 0.5 to 4 bar. Iritani and colleagues [108], in turn, while developing a simplified method for estimating fouling layer thickness in membrane filtration, reported a decrease in the porosity of the BSA fouling layer from 0.9 to 0.85 in the ultrafiltration process when the applied pressure was increased from 100 to 400 kPa. In a similar study on the influence of various operating conditions on membrane fouling, it was reported that the average cake porosity of a colloidal fouling layer formed on a membrane at a flux of 60 LMH was 0.7, but decreased to 0.4 when the flux was increased to 100 LMH [109]. This was attributed to the increased drag force, which had caused fouling layer compression by decreasing the voids between foulants. All other authors who investigated similar phenomena arrived at similar conclusions.

Existing studies also show that a porosity gradient always develops across fouling layers on a membrane surface during the filtration process. This is due to permeation drag. The porosity increases with distance from the membrane, as illustrated in Figure 2.8.

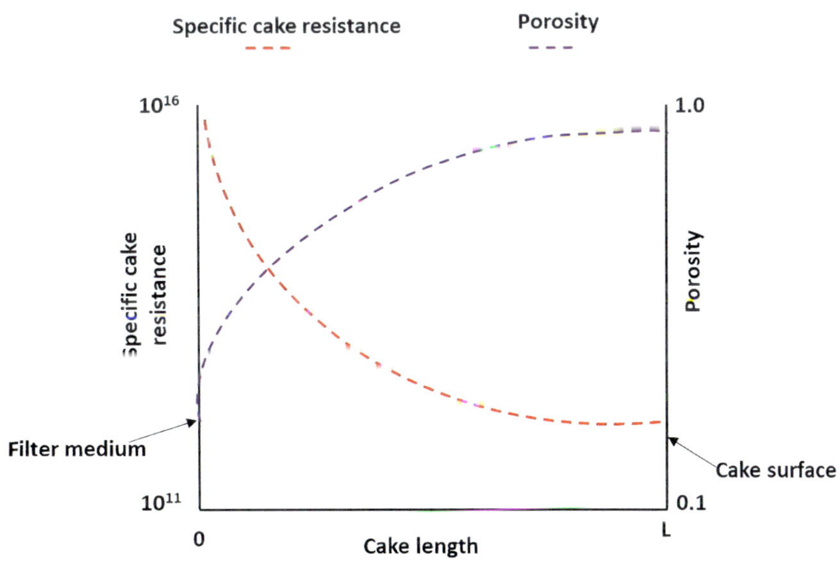

FIGURE 2.8 An illustration of the spatial distribution of porosity and specific cake resistance within a membrane fouling layer. Adapted with permission from Taylor & Francis [110].

In an investigation of the properties of a fouling layer in an immersed anaerobic membrane bioreactor, Gao and fellow researchers [111] compared the porosity at the top of a fouling layer to that at the bottom. The porosity increased from approximately 15% at the bottom to more than 45% towards the top. Jafari and colleagues [112] made similar observations during their evaluation of biofilm compressibility on an ultrafiltration membrane. They reported that the biofilm formed had a loose top layer and a dense base layer. The bottom layer had a porosity of approximately 0.15, whereas that at the top was approximately 0.8. The same spatial distribution trend of porosity within different microbial cakes formed on a membrane during the microfiltration process was reported by Katagiri's research group [110]. They found that as the compressive stresses approached zero at the surface of the fouling layer, the porosity increased to approximately 0.5 and 0.4 for the fouling layer of *E. coli* and baker's yeast, respectively. They fed the microbial filtration data from a concentration polarization (C-P) cell into a proposed empirical expression by Tiller and Leu, as shown in Equation (2.7) [113]:

$$\varepsilon = \varepsilon_0 \left(1 + \frac{P_s}{P_a}\right)^{\gamma} \tag{2.7}$$

where ε_0 and P_a are empirical constants. As a result, the researchers were able to relate the cake porosity (ε) to compressive stress (P_s) and concluded that the low cake porosity at the base of the fouling layer was a result of the high compressive stress near the membrane surface.

A decrease in fouling layer porosity due to an increase in foulant–foulant interactions with increasing permeation flux or applied pressure is reported to have a significant effect on the permeability and fouling performance of membrane filtration systems.

(c) Specific cake resistance/hydraulic resistance

It is well documented that fluid–particle interactions and, consequently, the compression of the fouling layer increase hydraulic resistance and, consequently, the specific cake resistance [96, 114, 115]. An increase in hydraulic resistance is associated with a decrease in the voidage and thickness of the fouling layer. Sioutopoulos and colleagues [116] studied the effect of permeation flux on fouling resistance in the ultrafiltration of a feed containing a mixture of organics. They observed an increase in specific cake resistance from 2.2×10^{15} to 5.5×10^{15} m/kg when the flux increased from 18 to 87 LMH. This increase was attributed to changes in the fouling layer structure as a result of fluid drag and the resultant compressive stresses. Jafari and co-workers [112] also evaluated and modelled the effect of hydrodynamics, i.e., permeation flow, on the specific cake resistance. They first developed a biofilm on a gravity-driven membrane for 20 days under 0.06 bar pressure and then subjected it to compression/relaxation through filtration. Their results showed that the fouling resistance increased from 1.2×10^{12} to 1.5×10^{12} m^{-1} when the applied pressure was increased from 0.06 to 0.5 bar. They explained that the variation in hydraulic resistance was linked to the change in the fouling layer thickness and porosity due to compression. The data of Karabelas and Sioutopoulos [117] from their fouling experiments of ultrafiltration and reverse osmosis membranes by an alginate solution also demonstrated the significant effect of fluid–particle dynamics on the specific cake resistance. Using the correlation presented in Equation (2.8), they found that specific cake resistance was strongly dependent on the permeation flux at a power greater than 1. Derlon and co-workers [100], in turn, evaluated the change in biofilm hydraulic resistance with increasing TMP. The resistance of the biofilm on an ultrafiltration membrane increased linearly with increasing TMP. Raising the pressure from 0.1 to 2.5 bar increased the biofilm resistance from 4×10^{12} to 1.6×10^{13} m^{-1}. From these studies, it is safe to conclude that permeation drag has a strong influence on the specific cake resistance and, consequently, on membrane fouling. This is mainly attributed to compression of the fouling layer:

$$\alpha = \varphi J^{\beta} \tag{2.8}$$

where α is the specific cake resistance, J is the flux, and φ and β are empirical constants.

Similar to porosity, the specific cake resistance is also reported to have a varied spatial distribution across the fouling layer. The decrease in porosity with decreasing distance from the membrane surface suggests an increase in the compactness of the fouling layer and, therefore, an increase in the specific cake resistance. Katagiri and team [110] evaluated the compressibility–permeability characteristics of microbial cake formed on a microfiltration membrane and found that the cake near the membrane exhibited a high resistance relative to the remaining part of the cake, as illustrated in Figure 2.8. They reported that this was due to the high solid compressive stress and, consequently, the compact nature of the bottom layer, which offered more resistance to flow.

2.3.2.4 Reversibility of Fouling Layer

As was discussed in Section 2.2.2, fouling can be classified based on the strength of attachment of foulant on a membrane. This affects its reversibility and consequently the cleaning performance of membrane systems. The strength of the attachment of foulants to membranes is affected by several factors, including permeation drag. Various researchers have reported that the ease of removal of the fouling layer decreases with increasing applied pressure or flux [20, 21, 118]. In their study of fouling and mitigation methods during the filtration of primary wastewater through a microfilter and an ultrafilter, Hube and co-workers [20] found that reversible and irreversible fouling rates were more pronounced at a high pressure of 50 kPa than at 8 kPa. In addition, they observed that a compact fouling layer that was difficult to remove with shear forces was formed at elevated pressure. This was attributed to an increase in foulant–foulant interactions and a resultant decrease in the inter-foulant distance. Loginov and co-workers [61] demonstrated the effect of operating pressure on the irreversible fouling of a membrane used in the ultrafiltration of a casein micelle suspension. They found that the flux recovery achieved by rinsing decreased as a function of TMP. The residual fouling layer post-rinsing increased from 4.9×10^{13} to 8.0×10^{13} m^{-1} when the applied pressure during filtration increased from 0.5 to 1.1 bar. The effect of fluid–particle dynamics due to permeation drag on fouling reversibility was evident from Daly and colleagues' [118] examination of two initial fluxes, 50 and 80 LMH. Even though similar fouling layer thicknesses of 174 μm and 179 μm were formed in the two separate runs, only 18% of the fouling layer could be removed through backwash for an initial flux of 80 LMH, compared to 72% for 50 LMH. The fouling layer formed at 80 LMH was found to be difficult to remove because of its density and strength as a result of the high permeation drag applied during the filtration process.

Some studies have further established that the irreversibility of fouling layers is only profound beyond a certain critical flux [119–121], called the critical flux of irreversibility. This is the flux below which reversible fouling occurs, but above which an irreversible fouling layer develops. This flux should be distinguished from the classic critical flux, which is normally defined as the flux below which little or no fouling occurs, but above which adverse fouling is observed. Xie and team [121] studied the use of critical flux to alleviate fouling of polyvinylidene fluoride (PVDF) -TiO_2 membranes and reported that no significant increase in the irreversible fouling rate was observed below a critical flux of 30 LMH at a TMP of approximately 60 kPa. However, when the operating TMP increased from 60 to 100 kPa, the irreversible fouling rate increased from 0.03×10^{12} m^{-1} to 0.82×10^{12} m^{-1}. The researchers concluded, therefore, that irreversible fouling could be limited by operating below the critical flux of irreversibility (or pressure). In another study, where the critical flux of irreversibility of a submerged ceramic membrane system during the harvesting of algal cells was identified to be around 35 LMH, irreversible fouling increased from approximately 1.6×10^{11} to 4.1×10^{11} m^{-1} when the flux increased from subcritical to supercritical levels [122]. The irreversible fouling deposited in the course of subcritical flux filtration represented 3% of the total fouling resistance, while that formed during supercritical flux was 10%. Nguyen's research group [120], in turn, observed 100% reversibility of an organic fouling layer formed on a forward osmosis membrane during filtration below a critical flux of irreversibility of approximately 30 LMH. However, when the filtration process was performed at the critical flux of 30 LMH and above (at 38 LMH), a fouling reversibility of approximately 97% and 85%, respectively, was observed with physical flushing. In their study, Diagne and co-workers [119] also reported that the amount of irreversible protein layer that remained on a PES ultrafiltration membrane after water rinsing doubled when the TMP used in filtering skimmed milk was increased above the critical pressure of irreversibility (1.5 bar) to 4 bar. From these studies, it can be concluded that the critical flux of irreversibility delimits two fouling reversibility behaviours of the layers formed on membranes: Below this critical flux, the fouling layer is mostly reversible; above the critical value, the layer becomes irreversible. The fouling layers formed above the critical flux appear to be more cohesive at high applied pressure, resulting in irreversible layers. This clearly shows the effect of fluid–particle dynamics due to permeation drag on the ease of physical removal of the membrane fouling layer, and the importance of identifying the critical flux of irreversibility of a membrane system to minimise irreversible fouling.

Other researchers have employed critical deposited mass or critical filtered volume to study the effects of fluid–particle dynamics on fouling layer reversibility [71, 123, 124]. In optimising the dead-end filtration conditions for an anoxic membrane bioreactor, McAdam and Judd identified the critical filtered volume to be 12.5 L/m^2. Beyond this volume, the reversibility of the fouling layer by backflushing and gas injection decreased [123]. Another study found that latex deposited on a hollow-fibre membrane was 10% and 20% less reversible by rinsing when the critical filtered volume was increased from 35 L/m^2 to 74 and 84 L/m^2, respectively [124]. Using Monte Carlo simulations and a force model equation, Guan and co-workers evaluated the effect of the deposited mass on the reversibility of the fouling layer in a dead-end membrane filtration system [71]. They compared their simulation results to those of previous experimental work and found that the percentage of fouling layer reversibility decreased with increasing deposited mass, as shown in Figure 2.8. The analysis by all the aforementioned authors suggests that there is a critical deposited mass or filtered volume (which, in turn, is affected by imposed flux) beyond which the membrane fouling layer becomes difficult to remove by physical means. As more layers form on top of an existing cake, the reference foulants down below develop an ever-repulsive force to the drag force caused by the newly added foulants above. However, when the drag force overcomes this repulsion, the reference foulants consolidate, making it difficult for the fouling layers to be removed, whether by rinsing, air scouring, or backwashing (Figure 2.9).

Several studies have also noted that the effects of fluid–particle dynamics on the reversibility of the fouling layer become more pronounced with filtration duration [125]. Cui and colleagues noted this in their investigation of an efficient backwash strategy for membrane bioreactors based on fouling layer compression theory. They reported that for a fouling layer formed at a flux of 140 LMH, the flux recovery ratio at a backwash flux of 210 LMH was 96.2% and 95.7% for filtration durations of 48 and 96 hours, respectively [125]. After filtration for 144 hours, however, the flux recovery rate decreased to 80%, and dropped even further to 44% after 174 hours. The researchers attributed this significant reduction in the recovery rate to the compression of the fouling layer with increasing filtration duration. In their study

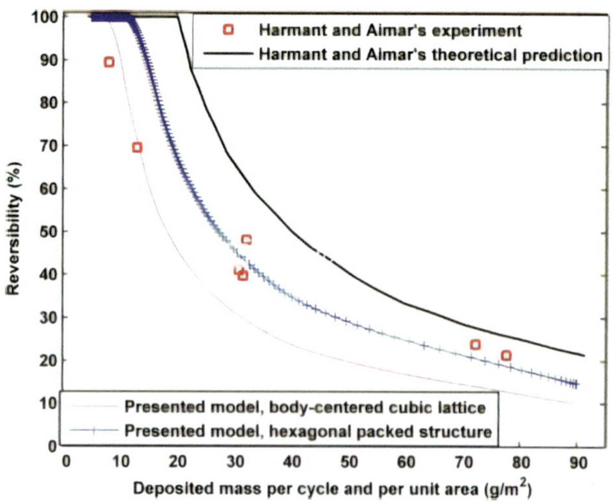

FIGURE 2.9 The reversibility of the membrane fouling layer as a function of mass of foulant deposited. Reused with permission from Elsevier B.V. [71].

of a membrane system operated at a subcritical flux, Cho and Fane [126] also reported a decrease in the reversibility of the fouling layer with a longer filtration duration. After each filtration run, the system was subjected to a relaxation period as a physical cleaning method. The post-relaxation residual pressure drop across the membrane was 4.5 kPa for a system operated for a shorter duration, and 12 kPa for a system operated for a longer duration. Therefore, the researchers concluded that membrane cleaning by relaxation was more effective when applied in the early stages of filtration. Another investigation revealed that the degree of fouling reversibility by backwash decreased from 93% to approximately 80% when the filtration duration exceeded 3600 seconds [127]. This was due to the formation of an irreversible compact fouling layer. Therefore, filtration duration does appear to contribute to the effects of fluid–particle interactions on the reversibility of membrane fouling.

2.4 CONCLUSION AND FUTURE PERSPECTIVES

This chapter reviews the role of the interplay of fluid dynamics and fluid–particle interactions on membrane fouling. These effects fall into four categories: the mass transfer and deposition of foulants onto the membrane, the compressibility of the fouling layer, a change in the structural properties of the fouling layer, and the reversibility of the fouling layer formed.

Greater permeation drag increases the amount of foulant deposited on the membrane and intensifies compression of the fouling layer. The compressibility, in turn, affects the structural properties of the fouling layer, increasing the cake density and specific cake resistance while decreasing porosity. Collectively, these factors lower the membrane permeability and render the fouling layer less reversible. Eventually, this affects the overall performance of membrane systems. Therefore, fluid–particle dynamics certainly play a critical role in the fouling and general performance of membrane systems.

Nevertheless, few studies have examined this crucial role of fluid–particle dynamics in respect of membrane systems that filter biological matter, such as membrane bioreactors (MBRs). As live biological matter such as bacteria is highly compressible, high permeation drag causes significant hydraulic compression, which results in high resistance to flow as well as the production of extracellular polymeric substances (EPS) by stressed bacteria. These outcomes affect the overall performance of MBRs and the lifespan of membranes. In filling the research gap relating to the impact of fluid–particle dynamics due to permeation drag on membrane systems that process live biological matter, investigations should specifically focus on the impact of permeation drag on the characteristics of the biofouling layer and, consequently, on the performance of MBRs.

Furthermore, there is no comprehensive study that accounts for all the effects of fluid–particle interactions on membrane fouling. To date, most investigations have focused on individual effects, which makes it difficult to evaluate how all the effects cumulatively contribute to membrane fouling and, consequently, influence the performance of membrane systems. What is required, therefore, is a model that relates the applied permeation drag, the compression of the fouling layer as well as the characteristics of the fouling layer to the performance of membrane systems (especially MBRs). This model can then be used to identify an optimum permeation drag regime for the operation of various membrane systems.

REFERENCES

1. A. Gul, J. Hruza, F. Yalcinkaya, Fouling and chemical cleaning of microfiltration membranes: A mini-review, *Polymers (Basel)* 13 (2021). https://doi.org/10.3390/polym13060846.
2. A.I. Cirillo, G. Tomaiuolo, S. Guido, Membrane fouling phenomena in microfluidic systems: From technical challenges to scientific opportunities, *Micromachines* 12 (2021). https://doi.org/10.3390/mi12070820.

3. S.A. Yala, V. Pillay, A predictive model for the effects of compression of fouling layers on the filtration performance of flat-sheet membrane bioreactors, *Desalin. Water Treat.* 323 (2025) 101383. https://doi.org/10.1016/j.dwt.2025.101383.
4. V. Sandoval-García, M.V. Ruano, M. Alliet, C. Brepols, J. Comas, J. Harmand, M. Heran, G. Mannina, I. Rodriguez-Roda, I. Smets, A. Robles, Modeling MBR fouling: A critical review analysis towards establishing a framework for good modeling practices, *Water Res.* 268 (2025) 122611. https://doi.org/10.1016/j.watres.2024.122611.
5. T. Wang, Z. Jin, Y. Yang, J. Ma, M. Aghbashlo, H. Zhang, S. Sun, M. Tabatabaei, J. Pan, In-depth insights into the temporal-based fouling mechanism and its exploration in anaerobic membrane bioreactors: A review, *J. Clean. Prod.* 375 (2022) 134110. https://doi.org/10.1016/j.jclepro.2022.134110.
6. Y. Xia Zhao, P. Li, R. Hong Li, X. Yan Li, Direct filtration for the treatment of the coagulated domestic sewage using flat-sheet ceramic membranes, *Chemosphere* 223 (2019) 383–390. https://doi.org/10.1016/j.chemosphere.2019.02.055.
7. W. Zheng, Y. Chen, X. Xu, X. Peng, Y. Niu, P. Xu, T. Li, Research on the factors influencing nanofiltration membrane fouling and the prediction of membrane fouling, *J. Water Process Eng.* 59 (2024) 104876. https://doi.org/10.1016/j.jwpe.2024.104876.
8. L. Böhm, A. Drews, H. Prieske, P.R. Bérubé, M. Kraume, The importance of fluid dynamics for MBR fouling mitigation, *Bioresour. Technol.* 122 (2012) 50–61. https://doi.org/10.1016/j.biortech.2012.05.069.
9. M. El Batouti, N.F. Alharby, M.M. Elewa, Review of new approaches for fouling mitigation in membrane separation processes in water treatment applications, *Separations* 9 (2022). https://doi.org/10.3390/separations9010001.
10. W. Zhang, J. Luo, L. Ding, M.Y. Jaffrin, A review on flux decline control strategies in pressure-driven membrane processes, *Ind. Eng. Chem. Res.* 54 (2015) 2843–2861. https://doi.org/10.1021/ie504848m.
11. T. Wen, Q. Huang, T. Fang, C. Xie, M. Li, W. Liu, K. Li, Fouling of reverse osmosis membrane in sugar mill condensate purification under sub- and super-boundary flux conditions, *J. Environ. Chem. Eng.* 12 (2024) 111974. https://doi.org/10.1016/j.jece.2024.111974.
12. S.A. Yala, V.L. Pillay, The effect of fluid dynamics and fluid-particle interactions on the characteristics of fouling layers in membrane filtration systems – A review, *J. Membr. Sci. Res.* 2029587 (2025) 1–14. https://doi.org/10.22079/jmsr.2024.2029587.1663.
13. Y.A. Tayeh, M.Y.D. Alazaiza, T.M. Alzghoul, M.J. Bashir, A comprehensive review of RO membrane fouling: Mechanisms, categories, cleaning methods and pretreatment technologies, *J. Hazard. Mater. Adv.* 18 (2025) 100684. https://doi.org/10.1016/j.hazadv.2025.100684.
14. X. Du, Y. Shi, V. Jegatheesan, I. Ul Haq, A review on the mechanism, impacts and control methods of membrane fouling in MBR system, 2020. https://doi.org/10.3390/membranes10020024.
15. S. Huang, R.H.A. Ras, X. Tian, Antifouling membranes for oily wastewater treatment: Interplay between wetting and membrane fouling, *Curr. Opin. Colloid Interface Sci.* 36 (2018) 90–109. https://doi.org/10.1016/j.cocis.2018.02.002.
16. P.K. Gkotsis, M.M. Mitrakas, A.K. Tolkou, A.I. Zouboulis, Batch and continuous dosing of conventional and composite coagulation agents for fouling control in a pilot-scale MBR, *Chem. Eng. J.* 311 (2017) 255–264. https://doi.org/10.1016/j.cej.2016.11.099.
17. N. AlSawaftah, W. Abuwatfa, N. Darwish, G.A. Husseini, A review on membrane biofouling: Prediction, characterization, and mitigation, *Membranes (Basel)* 12 (2022). https://doi.org/10.3390/membranes12121271.
18. H. Hamedi, O. Mohammadzadeh, S. Rasouli, S. Zendehboudi, A critical review of biomass kinetics and membrane filtration models for membrane bioreactor systems, *J. Environ. Chem. Eng.* 9 (2021) 106406. https://doi.org/10.1016/j.jece.2021.106406.
19. W. Guo, H.H. Ngo, J. Li, A mini-review on membrane fouling, *Bioresour. Technol.* 122 (2012) 27–34. https://doi.org/10.1016/j.biortech.2012.04.089.
20. S. Hube, J. Wang, L.N. Sim, D. Ólafsdóttir, T.H. Chong, B. Wu, Fouling and mitigation mechanisms during direct microfiltration and ultrafiltration of primary wastewater, *J. Water Process Eng.* 44 (2021) 102331. https://doi.org/10.1016/j.jwpe.2021.102331.
21. S. Shao, Y. Li, T. Jin, W. Liu, D. Shi, J. Wang, Y. Wang, Y. Jiang, J. Li, H. Li, Biofouling layer maintains low hydraulic resistances and high ammonia removal in the UF process operated at low flux, *J. Membr. Sci.* 596 (2020) 117612. https://doi.org/10.1016/j.memsci.2019.117612.
22. W. Yuan, X. Chen, Z. Yu, Y. Wan, J. Lin, W. Ye, Critical review of membrane fouling in reverse osmosis treatment: Characterizations, models, mechanisms, and controls, *Sep. Purif. Technol.* 363 (2025) 132119. https://doi.org/10.1016/J.SEPPUR.2025.132119.
23. A.L. Ahmad, N.F. Che Lah, S. Ismail, B.S. Ooi, Membrane antifouling methods and alternatives: Ultrasound approach, *Sep. Purif. Rev.* 41 (2012) 318–346. https://doi.org/10.1080/15422119.2011.617804.

24. F. Meng, S. Zhang, Y. Oh, Z. Zhou, H.S. Shin, S.R. Chae, Fouling in membrane bioreactors: An updated review, *Water Res.* 114 (2017) 151–180. https://doi.org/10.1016/j.watres.2017.02.006.
25. P.S. Goh, W.J. Lau, M.H.D. Othman, A.F. Ismail, Membrane fouling in desalination and its mitigation strategies, *Desalination* 425 (2018) 130–155. https://doi.org/10.1016/j.desal.2017.10.018.
26. A. Alkhatib, M.A. Ayari, A.H. Hawari, Fouling mitigation strategies for different foulants in membrane distillation, *Chem. Eng. Process. - Process Intensif.* 167 (2021) 108517. https://doi.org/10.1016/J.CEP.2021.108517.
27. B. Chamam, R. Ben Dassi, J. Abderraouf, J.P. Mericq, C. Faur, I. Trabelsi, L. El Mir, M. Heran, Incorporation of Ag-ZnO nanoparticles into PVDF membrane formulation to enhance dye retention, permeability, and antibacterial properties, *Polymers (Basel)* 17 (2025) 1–15. https://doi.org/10.3390/polym17091269.
28. F. Saffarimiandoab, B. Yavuzturk Gul, R.S. Tasdemir, S.E. Ilter, S. Unal, B. Tunaboylu, Y.Z. Menceloglu, İ. Koyuncu, A review on membrane fouling: Membrane modification, *Desalin. Water Treat.* 216 (2021) 47–70. https://doi.org/10.5004/dwt.2021.26815.
29. W. Song, D. Xu, X. Yang, J. Li, P. Zhao, X. Wang, H.Y. Ng, Membrane surface coated with zwitterions for fouling mitigation in membrane bioreactor: Performance and mechanism, *J. Membr. Sci.* 676 (2023) 121622. https://doi.org/10.1016/j.memsci.2023.121622.
30. M.K. Hadi, X. Wang, Y. Peng, S. Sangaraju, F. Ran, Functional polymeric membrane materials: A perspective from versatile methods and modification to potential applications, *Polym. Sci. Technol.* 1 (2025) 366–412. https://doi.org/10.1021/polymscitech.4c00030.
31. F. Matebese, R.M. Moutloali, Integrating ultrafiltration membranes with flocculation and activated carbon pretreatment processes for membrane fouling mitigation and metal ion removal from wastewater, *ACS Omega* 8 (2023) 9074–9085. https://doi.org/10.1021/acsomega.2c03524.
32. N. Morales, C. Mery-Araya, P. Guerra, R. Poblete, J. Chacana-Olivares, Mitigation of membrane fouling in membrane bioreactors using granular and powdered activated carbon: An experimental study, *Water (Switzerland)* 16 (2024). https://doi.org/10.3390/w16172556.
33. Y. Wang, M. Li, H. Yang, A review on mitigation of pressure-driven membrane fouling by coagulation/flocculation as a pretreatment in membrane separation processes, *Sep. Purif. Technol.* 364 (2025) 132370. https://doi.org/10.1016/J.SEPPUR.2025.132370.
34. J.D. Ladouceur, R.M. Narbaitz, C.Q. Lan, Phased-inline coagulation for low-pressure membranes in water and wastewater treatment: A review of fouling mitigation, process control, and water quality, *Environ. Technol. Rev.* 13 (2024) 305–324. https://doi.org/10.1080/21622515.2024.2343128.
35. K. Tian, X. Xu, J. Zhu, S. Cao, Z. Yin, F. Li, W. Yang, A critical review of oxidation for membrane fouling control in water treatment: Applications, mechanisms and challenges, *J. Environ. Chem. Eng.* 12 (2024) 114718. https://doi.org/10.1016/J.JECE.2024.114718.
36. Y. Xu, C. Luo, M. Yin, X. Cheng, X. Zhang, Y. Jin, R. Mu, J. Xu, K. Du, X. Jiang, S. Bai, D. Wu, X. Zhai, A novel pre-oxidation of ultrafiltration membrane fouling mitigation via ascorbic acid-activated hydrogen peroxide, *Chem. Eng. J.* 521 (2025) 166836. https://doi.org/10.1016/J.CEJ.2025.166836.
37. I. Ruigómez, E. González, L. Rodríguez-Gómez, L. Vera, Fouling control strategies for direct membrane ultrafiltration: Physical cleanings assisted by membrane rotational movement, *Chem. Eng. J.* 436 (2022). https://doi.org/10.1016/j.cej.2022.135161.
38. T. Xiao, Z. Zhu, L. Li, J. Shi, Z. Li, X. Zuo, Membrane fouling and cleaning strategies in microfiltration/ultrafiltration and dynamic membrane, *Sep. Purif. Technol.* 318 (2023) 123977. https://doi.org/10.1016/J.SEPPUR.2023.123977.
39. O. Díaz, L. Vera, E. González, Optimization of the operating parameters in a membrane bioreactor operated in direct-flow mode by assessing threshold flux for compressibility, *Chem. Eng. Sci.* 274 (2023) 118675. https://doi.org/10.1016/j.ces.2023.118675.
40. W. Zhang, W. Liang, J. Jin, S. Meng, Z. He, M. Ali, P.E. Saikaly, Filtration performance of biofilm membrane bioreactor: Fouling control by threshold flux operation, *Chemosphere* 362 (2024) 142458. https://doi.org/10.1016/j.chemosphere.2024.142458.
41. P. Le-Clech, V. Chen, T.A.G. Fane, Fouling in membrane bioreactors used in wastewater treatment, *J. Membr. Sci.* 284 (2006) 17–53. https://doi.org/10.1016/j.memsci.2006.08.019.
42. B. Díez, R. Rosal, A critical review of membrane modification techniques for fouling and biofouling control in pressure-driven membrane processes, *Nanotechnol. Environ. Eng.* 5 (2020). https://doi.org/10.1007/s41204-020-00077-x.
43. C. Lee, T.T. Nguyen, R.S. Adha, H.K. Shon, I.S. Kim, Influence of hydrodynamic operating conditions on organic fouling of spiral-wound forward osmosis membranes: Fouling-induced performance deterioration in FO-RO hybrid system, *Water Res.* 185 (2020) 116154. https://doi.org/10.1016/J.WATRES.2020.116154.

44. M. Ehsani, H. Doan, A. Lohi, A comprehensive review of membrane fouling and cleaning methods with emphasis on ultrasound-assisted fouling control processes, *Korean J. Chem. Eng.* 38 (2021) 1531–1555. https://doi.org/10.1007/s11814-021-0832-2.
45. S. Arefi-Oskoui, A. Khataee, M. Safarpour, Y. Orooji, V. Vatanpour, A review on the applications of ultrasonic technology in membrane bioreactors, *Ultrason. Sonochem.* 58 (2019) 104633. https://doi.org/10.1016/j.ultsonch.2019.104633.
46. H. Xiao, Z. Chen, J. Ding, N. Zhang, Z. Ye, Z. Xiao, S. Wang, P. Xie, Y. Chen, Effective and low-toxicity: A membrane cleaning method using peroxymonosulfate catalytic chlorination, *J. Hazard. Mater.* 462 (2024) 132827. https://doi.org/10.1016/J.JHAZMAT.2023.132827.
47. J. Lee, M. Zhan, Y. Kim, S. Hong, Comparison of different cleaning strategies on fouling mitigation in hollow fiber nanofiltration membranes for river water treatment, *J. Clean. Prod.* 380 (2022) 134764. https://doi.org/10.1016/j.jclepro.2022.134764.
48. T. Maliwan, J.H. Kho, J. Hu, Effective and low-impact backwashing strategies for an ultrafiltration membrane simultaneously fouled by microplastics and organic matter, *J. Clean. Prod.* 505 (2025) 145521. https://doi.org/10.1016/J.JCLEPRO.2025.145521.
49. N.R. Maddela, A.S. Abiodun, S. Zhang, R. Prasad, Biofouling in membrane bioreactors—Mitigation and current status: A review, *Appl. Biochem. Biotechnol.* (2022). https://doi.org/10.1007/s12010-022-04262-3.
50. K.L. Jepsen, M.V. Bram, L. Hansen, Z. Yang, S.M.Ø. Lauridsen, Online backwash optimization of membrane filtration for produced water treatment, *Membranes (Basel)* 9 (2019) 1–18. https://doi.org/10.3390/membranes9060068.
51. R. Habib, M.B. Asif, S. Iftekhar, Z. Khan, K. Gurung, V. Srivastava, M. Sillanpää, Influence of relaxation modes on membrane fouling in submerged membrane bioreactor for domestic wastewater treatment, *Chemosphere* 181 (2017) 19–25. https://doi.org/10.1016/J.CHEMOSPHERE.2017.04.048.
52. L. Fortunato, A.F. Lamprea, T.O. Leiknes, Evaluation of membrane fouling mitigation strategies in an algal membrane photobioreactor (AMPBR) treating secondary wastewater effluent, *Sci. Total Environ.* 708 (2020) 134548. https://doi.org/10.1016/J.SCITOTENV.2019.134548.
53. K. Gruskevica, L. Mezule, Cleaning methods for ceramic ultrafiltration membranes affected by organic fouling, *Membranes (Basel)* 11 (2021) 1–15. https://doi.org/10.3390/membranes11020131.
54. E.J. Sisay, A.N. Al-Tayawi, Z. László, S. Kertész, Recent advances in organic fouling control and mitigation strategies in membrane separation processes: A review, *Sustainability* 15 (2023) 13389. https://doi.org/10.3390/su151813389.
55. J.M.J. Millanar-Marfa, L. Borea, S.W. Hasan, M.D.G. de Luna, V. Belgiorno, V. Naddeo, Advanced membrane bioreactors for emerging contaminant removal and quorum sensing control, in: *Current Developments in Biotechnology and Bioengineering: Advanced Membrane Separation Processes for Sustainable Water and Wastewater Management — Case Studies and Sustainability Analysis*, 2020: pp. 117–147. https://doi.org/10.1016/B978-0-12-819854-4.00006-X.
56. S. Min, H. Lee, L. Deng, W. Guo, B. Xu, H. Yong Ng, C.T. Mehmood, Z. Zhong, R. Zamora, E. Khan, S. Ranjan Dash, J. Kim, M. Pishnamazi, P.K. Park, S.R. Chae, Advanced strategies for mitigation of membrane fouling in anaerobic membrane bioreactors for sustainable wastewater treatment, *Chem. Eng. J.* 485 (2024) 149996. https://doi.org/10.1016/j.cej.2024.149996.
57. B. Zang, H. Zhou, H. Nan, Y. Li, Q. Li, D. Sano, R. Chen, Exploring the potential positive impact of phage-bacterium interactions on membrane fouling mitigation, *Front. Environ. Sci. Eng.* 19 (2025). https://doi.org/10.1007/s11783-025-2059-7.
58. F. Zamani, A. Ullah, E. Akhondi, H.J. Tanudjaja, E.R. Cornelissen, A. Honciuc, A.G. Fane, J.W. Chew, Impact of the surface energy of particulate foulants on membrane fouling, *J. Memb. Sci.* 510 (2016) 101–111. https://doi.org/10.1016/j.memsci.2016.02.064.
59. T.-A. Nguyen, S. Yoshikawa, S. Ookawara, Steady state permeate flux estimation in cross-flow ultrafiltration of protein solution, *Sep. Sci. Technol.* 49 (2014) 1469–1478. https://doi.org/10.1080/01496395.2014.893533.
60. H. Yang, X. Yu, J. Liu, Z. Tang, T. Huang, Z. Wang, Y. Zhong, Z. Long, L. Wang, A concise review of theoretical models and numerical simulations of membrane fouling, *Water (Switzerland)* 14 (2022). https://doi.org/10.3390/w14213537.
61. M. Loginov, F. Doudiès, N. Hengl, M. Karrouch, N. Leconte, F. Garnier-Lambrouin, J. Pérez, F. Pignon, G. Gésan-Guiziou, On the reversibility of membrane fouling by deposits produced during crossflow ultrafiltration of casein micelle suspensions, *J. Memb. Sci.* 630 (2021) 119289. https://doi.org/10.1016/j.memsci.2021.119289.
62. J. Kim, F.A. DiGiano, Fouling models for low-pressure membrane systems, *Sep. Purif. Technol.* 68 (2009) 293–304. https://doi.org/10.1016/j.seppur.2009.05.018.

63. V. Chen, A.G. Fane, S. Madaeni, I.G. Wenten, Particle deposition during membrane filtration of colloids: Transition between concentration polarization and cake formation, *J. Memb. Sci.* 125 (1997) 109–122. https://doi.org/10.1016/S0376-7388(96)00187-1.
64. C.Y. Tang, T.H. Chong, A.G. Fane, Colloidal interactions and fouling of NF and RO membranes: A review, *Adv. Colloid Interface Sci.* 164 (2011) 126–143. https://doi.org/10.1016/j.cis.2010.10.007.
65. J. Liu, T. Huang, R. Ji, Z. Wang, C.Y. Tang, J.O. Leckie, Stochastic collision–attachment-based Monte Carlo simulation of colloidal fouling: Transition from foulant–clean-membrane interaction to foulant–fouled-membrane interaction, *Environ. Sci. Technol.* 54 (2020) 12703–12712. https://doi.org/10.1021/acs.est.0c04165.
66. C.Y. Tang, Y.N. Kwon, J.O. Leckie, The role of foulant–foulant electrostatic interaction on limiting flux for RO and NF membranes during humic acid fouling—Theoretical basis, experimental evidence, and AFM interaction force measurement, *J. Memb. Sci.* 326 (2009) 526–532. https://doi.org/10.1016/j.memsci.2008.10.043.
67. J. Liu, Z. Wang, C.Y. Tang, J.O. Leckie, Modeling dynamics of colloidal fouling of RO/NF membranes with a novel collision-attachment approach, *Environ. Sci. Technol.* 52 (2018) 1471–1478. https://doi.org/10.1021/acs.est.7b05598.
68. J. Liu, W. Fu, X. Yu, H. Yang, D. Zhao, Z. Wang, L. Wang, X. Li, C.Y. Tang, Relating critical and limiting fluxes to metastable and long-term stable fluxes in colloidal membrane filtration through collision-attachment theory, *Water Res.* 238 (2023) 120010. https://doi.org/10.1016/j.watres.2023.120010.
69. W. Sun, H. Zhou, X. Yu, D. Zhao, J. Liu, L. Chen, Z. Wang, G. Liu, Y. Qiu, Y. Hong, Collision-attachment simulation of membrane fouling by oppositely and similarly charged colloids, *Water Res.* 252 (2024) 121194. https://doi.org/10.1016/j.watres.2024.121194.
70. J. Liu, Y. Fan, Y. Sun, Z. Wang, D. Zhao, T. Li, B. Dong, C.Y. Tang, Modelling the critical roles of zeta potential and contact angle on colloidal fouling with a coupled XDLVO–collision attachment approach, *J. Memb. Sci.* 623 (2021) 119048. https://doi.org/10.1016/j.memsci.2021.119048.
71. K. Guan, Y. Liu, X. Yin, W. Zhu, Y. Chu, C. Peng, M. Lv, Q. Sun, P. Rao, J. Wu, Influence of operation conditions on cake structure in dead-end membrane filtration: Monte Carlo simulations and a force model, *Chem. Eng. Res. Des.* 124 (2017) 124–133. https://doi.org/10.1016/j.cherd.2017.06.008.
72. M. Xie, J. Lee, L.D. Nghiem, M. Elimelech, Role of pressure in organic fouling in forward osmosis and reverse osmosis, *J. Memb. Sci.* 493 (2015) 748–754. https://doi.org/10.1016/j.memsci.2015.07.033.
73. Y. Zhao, D. Lu, Y. Cao, S. Luo, Q. Zhao, M. Yang, C. Xu, J. Ma, Interaction analysis between gravity-driven ceramic membrane and smaller organic matter: Implications for retention and fouling mechanism in ultralow pressure-driven filtration system, *Environ. Sci. Technol.* 52 (2018) 13718–13727. https://doi.org/10.1021/acs.est.8b03618.
74. D. Zhao, L. Chen, M. Peng, B. Xue, Z. Yao, W. Huang, Z. Wang, J. Liu, The complex influence of membrane roughness on colloidal fouling: A dialectical perspective, *J. Memb. Sci.* 725 (2025) 124014. https://doi.org/10.1016/j.memsci.2025.124014.
75. J. Liu, Z. Tang, H. Yang, X. Li, X. Yu, Z. Wang, T. Huang, C.Y. Tang, Dissecting the role of membrane defects with low-energy barrier on fouling development through a collision attachment–Monte Carlo approach, *J. Memb. Sci.* 663 (2022) 120981. https://doi.org/10.1016/j.memsci.2022.120981.
76. S. Wang, X. Lu, L. Zhang, J. Guo, H. Zhang, Characterization of the initial fouling layer on the membrane surface in a membrane bioreactor: Effects of permeation drag, *Membranes (Basel)* 9 (2019). https://doi.org/10.3390/membranes9090121.
77. S.T. Kang, A. Subramani, E.M.V. Hoek, M.A. Deshusses, M.R. Matsumoto, Direct observation of biofouling in cross-flow microfiltration: Mechanisms of deposition and release, *J. Memb. Sci.* 244 (2004) 151–165. https://doi.org/10.1016/j.memsci.2004.07.011.
78. Y.P. Zhang, A.G. Fane, A.W.K. Law, Critical flux and particle deposition of bidisperse suspensions during crossflow microfiltration, *J. Memb. Sci.* 282 (2006) 189–197. https://doi.org/10.1016/j.memsci.2006.05.022.
79. X. Du, Z. Wang, L. Chao, A combined model for describing the trans-membrane pressure (P) variation in constant flux dead-end microfiltration (MF) process, *J. Environ. Chem. Eng.* 10 (2022) 107076. https://doi.org/10.1016/j.jece.2021.107076.
80. K. Arandia, N.K. Karna, T. Mattsson, H. Theliander, Monitoring membrane fouling using fluid dynamic gauging: Influence of feed characteristics and operating conditions, *Membranes (Basel)* 13 (2023) 1–26. https://doi.org/10.3390/membranes13100834.
81. S.A. Jones, Y.M.J. Chew, M.R. Bird, D.I. Wilson, The application of fluid dynamic gauging in the investigation of synthetic membrane fouling phenomena, *Food Bioprod. Process.* 88 (2010) 409–418. https://doi.org/10.1016/j.fbp.2010.07.004.
82. S. Jiang, S. Xiao, H. Chu, J. Sun, Z. Yu, W. Zhang, Y. Chen, X. Zhou, Y. Zhang, Performance enhancement and fouling alleviation by controlling transmembrane pressure in a vibration membrane system for algae separation, *J. Memb. Sci.* 647 (2022) 120252. https://doi.org/10.1016/j.memsci.2022.120252.

83. R. Jiraratananon, A. Chanachai, A study of fouling in the ultrafiltration of passion fruit juice, *J. Memb. Sci.* 111 (1996) 39–48. https://doi.org/10.1016/0376-7388(95)00270-7.
84. E.A. Baptista, P.C.R. Pinto, I.F. Mota, J.M. Loureiro, A.E. Rodrigues, Ultrafiltration of ethanol/water extract of *Eucalyptus globulus* bark: Resistance and cake build up analysis, *Sep. Purif. Technol.* 144 (2015) 256–266. https://doi.org/10.1016/j.seppur.2015.02.036.
85. L. Hou, Z. Wang, P. Song, A precise combined complete blocking and cake filtration model for describing the flux variation in membrane filtration process with BSA solution, *J. Memb. Sci.* 542 (2017) 186–194. https://doi.org/10.1016/j.memsci.2017.08.013.
86. H. Di, G.J.O. Martin, D.E. Dunstan, Characterization of particle deposition during crossflow filtration as influenced by permeate flux and crossflow velocity using a microfluidic filtration system, *Front. Chem. Sci. Eng.* 15 (2021) 552–561. https://doi.org/10.1007/s11705-020-1962-5.
87. Z. Wang, X. Du, H. Zhang, X. Wang, P. Song, K. Natsagdorj, B. Khan, R. Khurram, The application of error function for normalized flux prediction in dead-end microfiltration (MF) process, *Sep. Sci. Technol.* 56 (2021) 117–128. https://doi.org/10.1080/01496395.2019.1706572.
88. T. Mattsson, W.J.T. Lewis, Y.M.J. Chew, M.R. Bird, The use of fluid dynamic gauging in investigating the thickness and cohesive strength of cake fouling layers formed during cross-flow microfiltration, *Sep. Purif. Technol.* 198 (2018) 25–30. https://doi.org/10.1016/j.seppur.2017.01.040.
89. P. Desmond, K.T. Huisman, H. Sanawar, N.M. Farhat, J. Traber, E.O. Fridjonsson, M.L. Johns, H.C. Flemming, C. Picioreanu, J.S. Vrouwenvelder, Controlling the hydraulic resistance of membrane biofilms by engineering biofilm physical structure, *Water Res.* 210 (2022) 118031. https://doi.org/10.1016/j.watres.2021.118031.
90. C. Dreszer, J.S. Vrouwenvelder, A.H. Paulitsch-Fuchs, A. Zwijnenburg, J.C. Kruithof, H.C. Flemming, Hydraulic resistance of biofilms, *J. Memb. Sci.* 429 (2013) 436–447. https://doi.org/10.1016/j.memsci.2012.11.030.
91. N. Ghaffour, A. Qamar, Membrane fouling quantification by specific cake resistance and flux enhancement using helical cleaners, *Sep. Purif. Technol.* 239 (2020) 116587. https://doi.org/10.1016/j.seppur.2020.116587.
92. J. Liang, L. Yu, J. Wu, The dynamic change of specific cake resistance in membrane bioreactor due to periodical cake relaxation, *J. Environ. Chem. Eng.* 8 (2020) 103837. https://doi.org/10.1016/j.jece.2020.103837.
93. E. Poorasgari, T.V. Bugge, M.L. Christensen, M.K. Jørgensen, Compressibility of fouling layers in membrane bioreactors, *J. Memb. Sci.* 475 (2015) 65–70. https://doi.org/10.1016/j.memsci.2014.09.056.
94. M.K. Jørgensen, T.V. Bugge, M.L. Christensen, K. Keiding, Modeling approach to determine cake buildup and compression in a high-shear membrane bioreactor, *J. Memb. Sci.* 409–410 (2012) 335–345. https://doi.org/10.1016/j.memsci.2012.04.002.
95. C.M. Chew, M.K. Aroua, M.A. Hussain, A practical hybrid modelling approach for the prediction of potential fouling parameters in ultrafiltration membrane water treatment plant, *J. Ind. Eng. Chem.* 45 (2017) 145–155. https://doi.org/10.1016/j.jiec.2016.09.017.
96. M.K. Amosa, M.S. Jami, M.F.R. Alkhatib, T. Majozi, S.A. Abdulkareem, Cake compressibility analysis of BPOME from a hybrid adsorption-microfiltration process, *Water Environ. Res.* 89 (2017) 292–300. https://doi.org/10.2175/106143017x14839994522948.
97. X. Du, F. Qu, H. Liang, K. Li, H. Chang, G. Li, Cake properties in ultrafiltration of TiO_2 fine particles combined with HA: in situ measurement of cake thickness by fluid dynamic gauging and CFD calculation of imposed shear stress for cake controlling, *Environ. Sci. Pollut. Res.* 23 (2016) 8806–8818. https://doi.org/10.1007/s11356-015-5984-3.
98. T.V. Bugge, M.K. Jørgensen, M.L. Christensen, K. Keiding, Modeling cake buildup under TMP-step filtration in a membrane bioreactor: Cake compressibility is significant, *Water Res.* 46 (2012) 4330–4338. https://doi.org/10.1016/j.watres.2012.06.015.
99. A.R. Costa, M. Norberta de Pinho, M. Elimelech, Mechanisms of colloidal natural organic matter fouling in ultrafiltration, *J. Memb. Sci.* 281 (2006) 716–725. https://doi.org/10.1016/j.memsci.2006.04.044.
100. N. Derlon, A. Grütter, F. Brandenberger, A. Sutter, U. Kuhlicke, T.R. Neu, E. Morgenroth, The composition and compression of biofilms developed on ultrafiltration membranes determine hydraulic biofilm resistance, *Water Res.* 102 (2016) 63–72. https://doi.org/10.1016/j.watres.2016.06.019.
101. A. Robles, M. V. Ruano, J. Ribes, A. Seco, J. Ferrer, Mathematical modelling of filtration in submerged anaerobic MBRs (SAnMBRs): Long-term validation, *J. Memb. Sci.* 446 (2013) 303–309. https://doi.org/10.1016/j.memsci.2013.07.001.
102. C. Dreszer, A.D. Wexler, S. Drusová, T. Overdijk, A. Zwijnenburg, H.C. Flemming, J.C. Kruithof, J.S. Vrouwenvelder, In-situ biofilm characterization in membrane systems using Optical Coherence Tomography: Formation, structure, detachment and impact of flux change, *Water Res.* 67 (2014) 243–254. https://doi.org/10.1016/j.watres.2014.09.006.

103. S.J. Skinner, A.D. Stickland, P.J. Scales, Predicting transmembrane pressure rise from biofouling layer compressibility and permeability, *Chem. Eng. Technol.* 41 (2018) 51–60. https://doi.org/10.1002/ceat.201700117.
104. C. Wang, W.N. Chen, Q.Y. Hu, M. Ji, X. Gao, Dynamic fouling behavior and cake layer structure changes in nonwoven membrane bioreactor for bath wastewater treatment, *Chem. Eng. J.* 264 (2015) 462–469. https://doi.org/10.1016/j.cej.2014.11.132.
105. R. Valladares Linares, A.D. Wexler, S.S. Bucs, C. Dreszer, A. Zwijnenburg, H.C. Flemming, J.C. Kruithof, J.S. Vrouwenvelder, Compaction and relaxation of biofilms, *Desalin. Water Treat.* 57 (2016) 12902–12914. https://doi.org/10.1080/19443994.2015.1057036.
106. C. Sun, L. Fiksdal, A. Hanssen-Bauer, M.B. Rye, T.O. Leiknes, Characterization of membrane biofouling at different operating conditions (flux) in drinking water treatment using confocal laser scanning microscopy (CLSM) and image analysis, *J. Memb. Sci.* 382 (2011) 194–201. https://doi.org/10.1016/j.memsci.2011.08.010.
107. M. Abbasi, R.S. Mohammad, A. Salahi, B. Mirza, Modeling of membrane fouling and flux decline in microfiltration of oily wastewater using ceramic membranes, *Chem. Eng. Commun.* 199 (2012) 78–93. https://doi.org/10.1080/00986445.2011.570391.
108. E. Iritani, N. Katagiri, H. Masuda, Simplified estimate of cake porosity in dead-end ultrafiltration of protein solution, *J. Chem. Eng. Japan* 51 (2018) 589–595. https://doi.org/10.1252/jcej.18we033.
109. S.T.V. Sim, A.H. Taheri, T.H. Chong, W.B. Krantz, A.G. Fane, Colloidal metastability and membrane fouling - Effects of crossflow velocity, flux, salinity and colloid concentration, *J. Memb. Sci.* 469 (2014) 174–187. https://doi.org/10.1016/j.memsci.2014.06.020.
110. N. Katagiri, H. Kawahara, Y. Arai, E. Iritani, Evaluation of compression-permeability characteristics of microbial cake based on microfiltration data, *Sep. Sci. Technol.* 51 (2016) 845–852. https://doi.org/10.1080/01496395.2015.1119849.
111. W.J. Gao, H.J. Lin, K.T. Leung, H. Schraft, B.Q. Liao, Structure of cake layer in a submerged anaerobic membrane bioreactor, *J. Memb. Sci.* 374 (2011) 110–120. https://doi.org/10.1016/j.memsci.2011.03.019.
112. M. Jafari, N. Derlon, P. Desmond, M.C.M. van Loosdrecht, E. Morgenroth, C. Picioreanu, Biofilm compressibility in ultrafiltration: A relation between biofilm morphology, mechanics and hydraulic resistance, *Water Res.* 157 (2019) 335–345. https://doi.org/10.1016/j.watres.2019.03.073.
113. F.M. Tiller, W.F. Leu, Basic data fitting in filtration, *J. Chinese Inst. Chem. Eng.* 11 (1980) 61–70.
114. N. Katagiri, Filterability and cake compressibility in dead-end membrane filtration controlled by cake formation, *Filtration* 17 (2017) 108–120.
115. A.J. Karabelas, D.C. Sioutopoulos, New insights into organic gel fouling of reverse osmosis desalination membranes, *Desalination* 368 (2015) 114–126. https://doi.org/10.1016/j.desal.2015.01.029.
116. D. Sioutopoulos, A. Karabelas, V. Mappas, Membrane fouling due to protein—polysaccharide mixtures in dead-end ultrafiltration; the effect of permeation flux on fouling resistance, *Membranes (Basel)* 9 (2019) 6–10. https://doi.org/10.3390/membranes9020021.
117. A.J. Karabelas, D.C. Sioutopoulos, Toward improvement of methods for predicting fouling of desalination membranes - The effect of permeate flux on specific fouling resistance, *Desalination* 343 (2014) 97–105. https://doi.org/10.1016/j.desal.2013.10.008.
118. S. Daly, A. Allen, V. Koutsos, A.J.C. Semião, Influence of organic fouling layer characteristics and osmotic backwashing conditions on cleaning efficiency of RO membranes, *J. Memb. Sci.* 616 (2020) 118604. https://doi.org/10.1016/j.memsci.2020.118604.
119. N.W. Diagne, M. Rabiller-Baudry, L. Paugam, On the actual cleanability of polyethersulfone membrane fouled by proteins at critical or limiting flux, *J. Memb. Sci.* 425–426 (2013) 40–47. https://doi.org/10.1016/j.memsci.2012.09.001.
120. T.T. Nguyen, C. Lee, R.W. Field, I.S. Kim, Insight into organic fouling behavior in polyamide thin-film composite forward osmosis membrane: Critical flux and its impact on the economics of water reclamation, *J. Memb. Sci.* 606 (2020) 118118. https://doi.org/10.1016/j.memsci.2020.118118.
121. W. Xie, J. Li, F. Sun, W. Dong, Z. Dong, Strategy study of critical flux/threshold flux on alleviating protein fouling of PVDF-TiO2 modified membrane, *J. Environ. Chem. Eng.* 9 (2021) 106148. https://doi.org/10.1016/j.jece.2021.106148.
122. X. Wu, C. Zhou, K. Li, W. Zhang, Y. Tao, Probing the fouling process and mechanisms of submerged ceramic membrane ultrafiltration during algal harvesting under sub- and super-critical fluxes, *Sep. Purif. Technol.* 195 (2018) 199–207. https://doi.org/10.1016/j.seppur.2017.12.001.
123. E.J. McAdam, S.J. Judd, Optimisation of dead-end filtration conditions for an immersed anoxic membrane bioreactor, *J. Memb. Sci.* 325 (2008) 940–946. https://doi.org/10.1016/j.memsci.2008.09.032.
124. Y. Bessiere, N. Abidine, P. Bacchin, Low fouling conditions in dead-end filtration: Evidence for a critical filtered volume and interpretation using critical osmotic pressure, *J. Memb. Sci.* 264 (2005) 37–47. https://doi.org/10.1016/j.memsci.2005.04.018.

125. Z. Cui, X. Wang, H. Ngo, G. Zhu, In-situ monitoring of membrane fouling migration and compression mechanism with improved ultraviolet technique in membrane bioreactors, *Bioresour. Technol.* 347 (2022) 126684. https://doi.org/10.1016/j.biortech.2022.126684.
126. B.D. Cho, A.G. Fane, Fouling transients in nominally sub-critical flux operation of a membrane bioreactor, *J. Memb. Sci.* 209 (2002) 391–403. https://doi.org/10.1016/S0376-7388(02)00321-6.
127. Y. Ye, L.N. Sim, B. Herulah, V. Chen, A.G. Fane, Effects of operating conditions on submerged hollow fibre membrane systems used as pre-treatment for seawater reverse osmosis, *J. Memb. Sci.* 365 (2010) 78–88. https://doi.org/10.1016/j.memsci.2010.08.038.

Microprocesses at Solid–Liquid Interface in Fluid–Particle Systems

3

Benjamin Oyegbile

3.1 INTRODUCTION TO FLUID–PARTICLE MIXING

Hydrodynamic-mediated interactions often occur in many technical and natural environmental processes. In the case of turbulent flows, this leads to the formation of turbulent eddies of various scales and sizes [1, 2]. These energy-carrying eddies often interact with particles and biological materials on various temporal and spatial scales. Eddy–particle interaction often plays a crucial role in these processes, and it is often the dominant driver of mass and momentum transfer. In studying the dynamics of such complex interplay of forces, a good knowledge of vortex dynamics and its influence on fluid and particle dynamics is highly indispensable [3, 4].

Turbulent mixing, particle dispersion, and bioreactions have been topics of intense and sustained interest in many scientific inquiries [5, 6]. The role of mixing and turbulence-driven particle dispersion in many fluid–particle processes is well understood owing to an abundant body of knowledge from many scientific interrogations. However, turbulence as a phenomenon is still poorly understood due to its complex nature. Since mixing and chemical reactions are impacted by the presence of turbulence, it is therefore extremely important to understand the different scales of turbulence in mixing applications.

As mentioned earlier, turbulence has been shown to lead to the formation of eddies on different scales [7, 8]. The spatial degree of mixing such as macromixing, mesomixing, and micromixing is governed by the different scales of turbulence [5]. Mixing, especially at high Reynolds numbers, is often characterised by irregular, rotational, and dissipative motion containing vorticities of different energy spectra or eddy sizes [3]. It is therefore imperative to carry out qualitative and quantitative assessments of mixing efficiency in many of the practical applications involving mixing and dispersion. A number of techniques are available for quantifying mixing performance in a wide range of applications. One widely used parameter for quantifying the mixing performance is the coefficient of variation proposed by Alloca and Streiff [6]. This approach relies on the statistical analysis of the spatiotemporal homogeneity of the particle dispersion in mixing applications [9]. Several other techniques are also available for quantifying the degree of

DOI: 10.1201/9781003518426-3

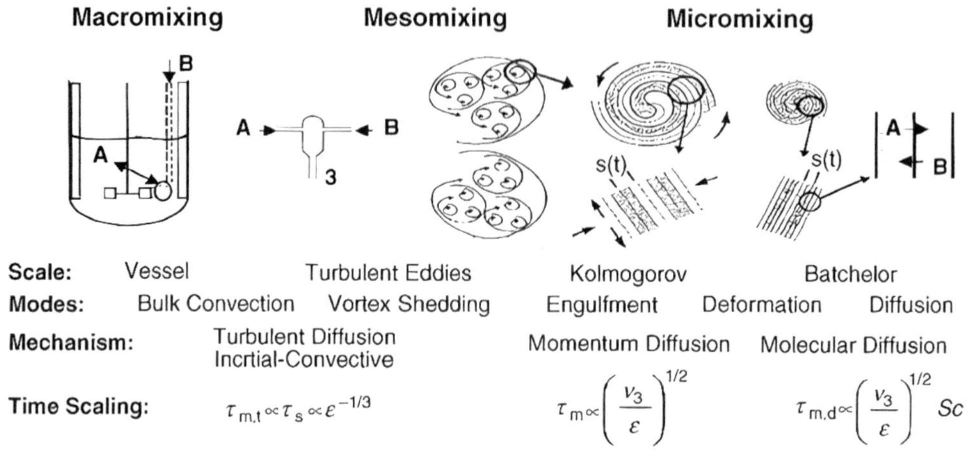

FIGURE 3.1 Schematic representation of different mixing and transport mechanisms: (a) bulk mixing or advection, (b) turbulent transport, and (c) molecular diffusion. Reprinted with permission from John Wiley [10].

mixing in bioreactor systems. In terms of the different mechanisms responsible for fluid–particle mixing, advective, turbulent, and diffusive transport depicted in Figure 3.1 are the dominant ones.

In turbulent mixing, energy transfer occurs on different eddy scales or energy spectrum. These turbulent eddies consist of the large energy-carrying eddies at the inertia sub-range to the smallest ones at the dissipation sub-range as shown in Figure 3.2. The important scales of energy spectra with respect to the different mixing regimes (i.e., micromixing, mesomixing, and macromixing) are the Taylor, Batchelor, and Kolmogorov length scales (Equations 3.1–3.3). The Kolmogorov length scale of turbulence is used as a convenient reference point for comparison of different scales of mixing. A detailed description of the different time and length scales in turbulent mixing applications is beyond the scope of this work. Further discussion on the subject matter can be found in the following reference texts [11, 12]. Therefore, getting the desired mixing regime is highly imperative for enhanced mass and momentum transfer.

In fluid–particle mixing, there exist three distinct mixing regimes in most practical mixing applications, namely micromixing, mesomixing, and macromixing. In typical mixing conditions, the dividing line between

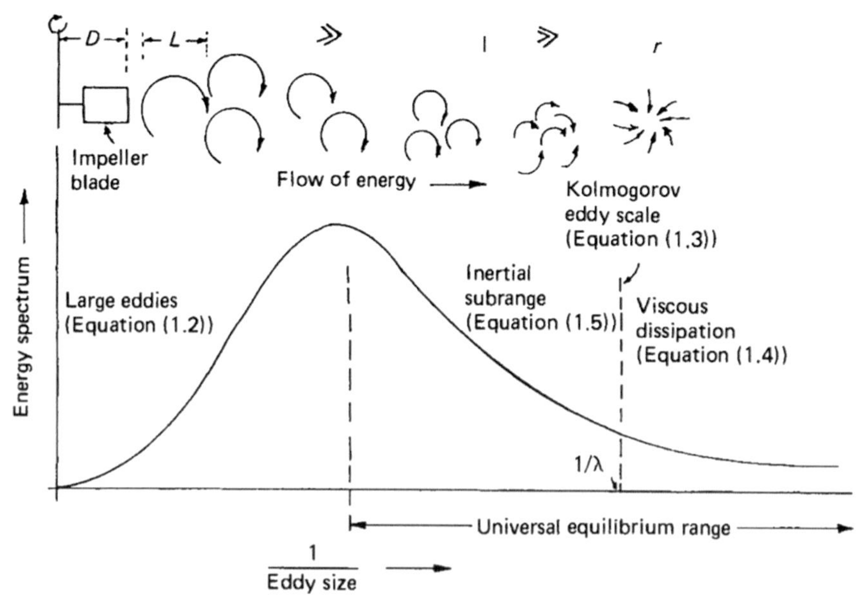

FIGURE 3.2 Schematic view of energy spectrum in turbulent mixing. Reproduced with permission from Elsevier B.V. [26].

micro- and macroscale is between 100 and 1000 μm, respectively [13]. Consequently, it is often necessary to tailor the mixing performance of physical, chemical, and biological processes in the target reactor systems to the expected mixing regime. Macromixing is largely driven by the largest scale of motion in the fluid or the integral length scale. Mesomixing on the other hand involves mixing on a smaller scale than the bulk circulation, but larger than the micromixing, while micromixing refers to the mixing on the smallest scale of fluid motion or molecular level. The largest eddies in turbulent dispersion that represent the macroscale of turbulence is produced by the stirrer or the agitator head and contains most of the fluid energy [11, 14].

Turbulent flow can be viewed as an eddy continuum, with different sizes ranging from the dimension of the turbulence-generating device to the Kolmogorov length scale. In between the energy-containing and energy-dissipating eddies, there exist many eddies of other scales smaller than the integral scale that continually transfer the kinetic energy of the fluid through the other length scales. The Taylor and Batchelor scales given in Equations (3.1) and (3.2) are examples of other important length scales of fluid motion. The Taylor scale is an intermediate length scale in the viscous sub-range that is representative of the energy transfer from large to small scales, but not a dissipation scale and does not represent any distinct group of eddies. Batchelor scale on the other hand is a limiting length scale where the rate of molecular diffusion is equal to the rate of dissipation of turbulent kinetic energy of the fluid [15, 16].

$$\lambda_0 = \left[\frac{v^2}{\varepsilon}\right]^{\frac{1}{4}} \tag{3.1}$$

$$\lambda_B = \left[\frac{D^2 v}{\varepsilon}\right]^{\frac{1}{4}} \tag{3.2}$$

$$\lambda_T = \frac{u\sqrt{15}}{\sqrt{\frac{\varepsilon}{v}}} \tag{3.3}$$

3.2 COLLOIDAL STABILITY AND INTERACTION FORCES

Colloidal materials in environmental and biological systems consist of small particles with a very large surface area. Their typical sizes are in the range of 0.001 to 10 μm, as shown in Table 3.1. The stability of these colloidal materials when dispersed in fluid can be explained by their tendency to acquire electrostatic charges by adsorbing ions from their surroundings. Forces mediating particle–particle interactions can broadly be classified into the following categories depending on the chosen approach: contact forces due to the particle–particle collisions and non-contact forces due to molecular interactions at the contact or interface. A brief description of these forces is hereby presented in the following sections.

3.2.1 Molecular Dynamics Particle Force Description

Molecular dynamics description of particle–particle interaction forces shown in Figure 3.3 is based on the molecular interactions through interfacial forces on the surface of the interacting particles [17]. Interfacial forces are generally assumed to act on a length scale smaller than the particle size, and interactions are only possible when particles are in close proximity or during collisions. The interactions of particles in suspension depend on these surface forces, which consist of the long-range ionic electrostatic repulsive forces and the short-range London–van der Waals attractive forces. Classical Derjaguin–Landau–Verwey–Overbeek

TABLE 3.1 Classification of Particles in Dispersion [16]

PARTICLE SIZES (MM)	*CLASSIFICATION*	*EXAMPLES*	*TOTAL SURFACE AREA (M^2CM^{-3})*	*TIME REQUIRED TO SETTLE 100 MM IF SG = 2.65*
$10–10^{-1}$	Coarse dispersion	Sand, mineral substances, precipitated and flocculated substances, silt, microplankton	$6\times10^{-4}–6\times10^{2}$	0.1–13 s
$10^{-2}–10^{-4}$	Fine particle dispersion	Mineral substances, precipitated and flocculated substances, silt, bacteria, plankton, and other microorganisms	0.6–60	11 min–2 years
$10^{-5}–10^{-6}$	Colloidal dispersion	Mineral substances, hydrolysed and precipitated products, macromolecules, biopolymers, viruses	6×10^{3}	20 years
$<10^{-6}$	Solution	Inorganic simple and complex ions, molecules and polymeric species, polyelectrolytes, organic molecules, and undissociated solutes	—	—

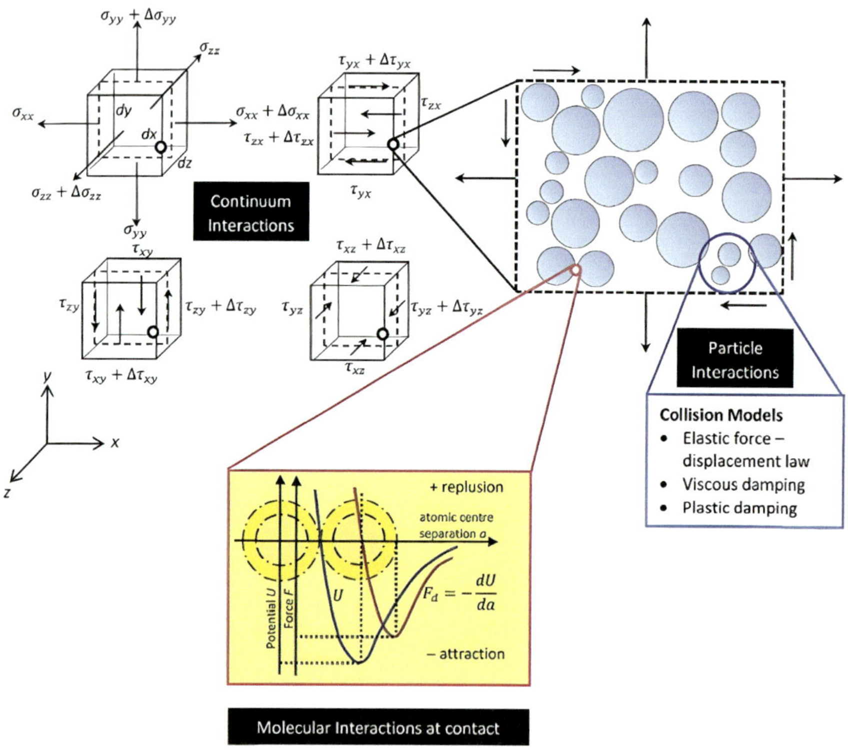

FIGURE 3.3 Classification of different phenomena in fluid–particle interactions. Reproduced with permission from Elsevier B.V. [17].

(DLVO) or colloid stability theory provides a quantitative description of the forces experienced by particles in close proximity by considering such interaction forces to be additive [18].

The magnitude of the electrostatic repulsive force due to the electric double layer and the London–van der Waals attractive force between two charged particles can be derived from their corresponding interaction potential energies [17]. Equations (3.4) and (3.5) represent the magnitude of these forces while Equations (3.6) and (3.7) represent their respective potential energies. In addition to these surface forces, a number of other interfacial interactions such as hydration effects, hydrophobic attraction, steric repulsion, and polymer bridging have been observed to mediate particle–particle interactions [19]. Additional forces due to the fluid–particle interactions must also be considered to fully resolve all the forces experienced by particles in suspension, and this is briefly discussed in Section 3.3.

$$F_{vdw} = \frac{dU_{vdw}}{dR_{ij}} \tag{3.4}$$

$$F_{elec} = \frac{dU_{elec}}{dR_{ij}} \tag{3.5}$$

$$U_{vdw} = \frac{A_H}{6}\left[\frac{2R_i^2}{R_{ij}^2 - 4R_i^2} + \frac{2R_i^2}{R_{ij}^2} + \ln\left\{1 - \frac{4R_i^2}{R_{ij}^2}\right\}\right] \tag{3.6}$$

$$U_{elec} = 2\pi\varepsilon R_i \psi_0^2 \ln\left[1 + \exp\left(-K\left(R_{ij} - 2R_i\right)\right)\right] \tag{3.7}$$

where R_{ij} is the distance between two interacting particles (centre-to-centre), A_H is the Hamaker constant, ε is the permittivity of the medium, ψ_0 is the surface potential of the particles, and K is the reciprocal of the Debye length.

3.2.2 Micromechanical Particle Force Description

In contrast to the molecular dynamics approach, the micromechanical description of particle–particle interaction relies on the geometric analysis of a finite number of discrete sub-elements, as shown in Figure 3.3. All particle–particle interactions within this context are described by contact forces in the normal and tangential directions while considering the elastic force–displacement, inelastic deformations or plastic dislocations, solid friction, and viscous damping [17]. Based on the simplicity of the interactions, a soft or hard sphere description can be given. In the hard sphere model, only elastic force–displacement is allowed. Soft sphere model, on the other hand, allows for most of the interactions that are possible when two particles are in direct contact.

When all these contact forces are fully resolved, the behaviour of the particles upon collision with another particle or impact on a wall, such as their translational and rotational velocities, can be predicted with a high degree of accuracy. A detailed theoretical description of the micromechanical theory of particle collisions and its importance in the determination of particle trajectory in dispersed systems is beyond the scope of this communication and is available elsewhere [20]. In addition to the contact forces, body forces such as gravity, and buoyancy and surface forces due to the fluid are some of the other important forces acting on the particles, and their quantification is highly indispensable in resolving the dynamics of particles in suspension [21]. Some of these additional forces are discussed in the next section.

3.3 PARTICLE–VORTEX INTERACTIONS IN TURBULENCE

In order to initiate meaningful interactions through the interfacial forces, particles must be brought in close proximity. This can be achieved through the mechanisms of Brownian motion, differential settling, or turbulent dispersion, as shown in Figure 3.4. The probability of particle collisions and the frequency of such collisions also depend on the trajectory of the particle motion. Thomas et al. [22] identified two types

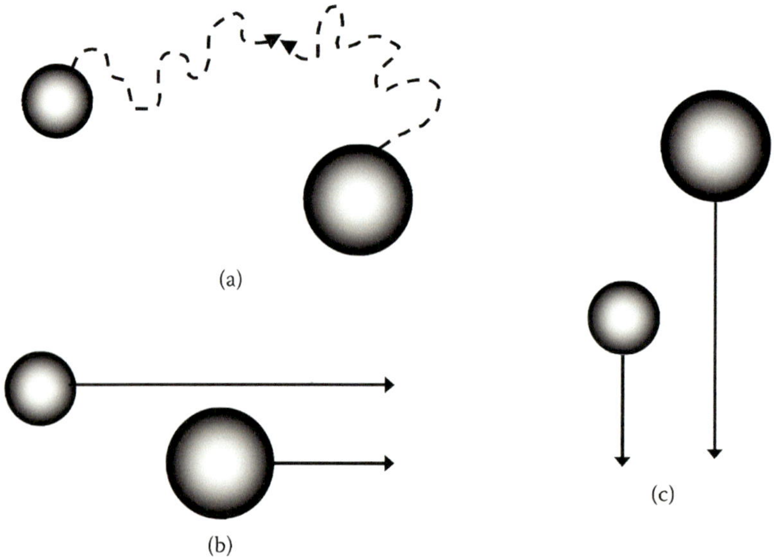

FIGURE 3.4 Mechanisms of particle transport in fluid leading to collisions: (a) Brownian motion, (b) fluid turbulence, and (c) differential settling. Reprinted with permission from CRC Press [19].

of particle trajectories leading to particle collisions, namely curvilinear and rectilinear particle trajectories. The particles in suspension under the influence of turbulence will experience fluctuating fluid motion with the particle being transported by the fluid eddies that exist within the flow vortex [3].

Consequently, small particles suspended in fluid exist in an environment of small energy-dissipating eddies in most practical flow devices. Under such conditions, particle collisions are facilitated by eddy size similar to that of the colliding particles. Furthermore, studies have shown that in addition to the fluid properties, the particle properties such as size and density also play an important role in particle–vortex interactions [1, 23]. Smaller particles with lower density than the flow fluid have been shown to be fully entrained within the vortex in the case of a vortical flow produced by the interaction of two flow streams of unequal velocity. The particles will be in dynamic equilibrium with the carrier fluid and will faithfully follow the flow streamlines. On the other hand, large particles will be unaffected by the vortex due to their large inertia while the intermediate particles will be driven from the centre of the vortex to the periphery, as shown in Figure 3.5. The determining factor in particle–vortex interactions is the ratio of the particle relaxation time to that of the fluid, which is given by the Stokes number (Equation 3.8). Depending on the flow scenario, the particles in suspension will experience additional forces such as the drag, lift, pressure gradient, virtual mass, Basset, and viscous stress forces due to the fluid–particle interactions [4, 24]. Taken together, all these forces will ultimately determine the trajectory, dynamics, and fate of particles in suspension. The trajectory equation is given in Equation (3.9):

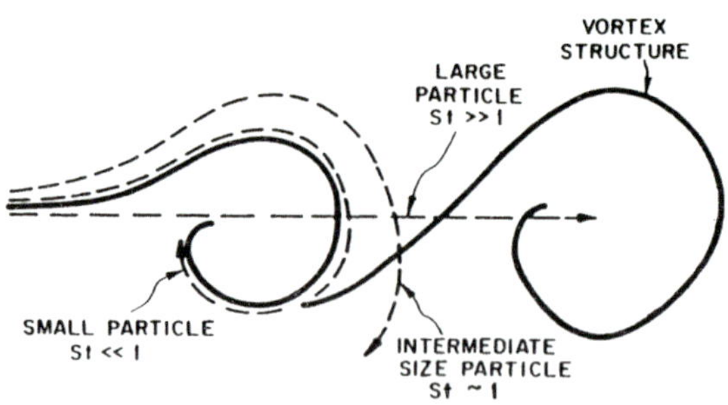

FIGURE 3.5 Vortex–particle interactions in turbulent flow. Reproduced with permission from Springer [1].

$$St = \frac{\tau_A}{\tau_F} = \frac{\varrho_p d^2 U}{18\mu L} \tag{3.8}$$

where τ_A is the particle relaxation time, τ_F is the time associated with fluid motion (fluid time), ϱ_p is the particle density, d is the particle diameter, L is the length scale associated with the vortex, and U is the flow velocity.

$$m\frac{d\vec{v}}{dt} = \overrightarrow{F_{mol}} + \overrightarrow{F_{mic}} + \overrightarrow{F_{f\text{-}p}} \tag{3.9}$$

3.4 VORTEX DYNAMICS AND AGGREGATE DISRUPTION

Turbulence is the main driver of particle interactions in many practical applications. Consequently, the vortex dynamics in terms of the particle collisions, coalescence aggregation, growth, and breakage is primarily controlled by the fluid turbulence. The aggregate stability under the influence of hydrodynamic force has been suggested to be a function of the binding or cohesive force F_B and the hydrodynamic breaking force F_H. While the binding force is determined by the aggregate structure and physicochemical attributes discussed in Section 3.2, flow turbulence is the principal factor in the case of the hydrodynamic force. Therefore, the dynamics of particle behaviour depends on an interplay of collision-induced particle aggregation and cohesive force and the rate of aggregate breakage due to the hydrodynamic stress.

The aggregate cohesive strength τ is a function of the physicochemical and biological conditions as well as the aggregate properties, while the hydrodynamic stress σ depends on the design of the aggregation unit and the prevailing process conditions. Several empirical and theoretical formulations are available for predicting the aggregate cohesive force and the maximum hydrodynamic breaking force. The global hydrodynamic stress σ due to the shearing action of the fluid motion on the aggregate as well as the overall cohesive strength τ of the aggregate resisting the hydrodynamic loading, assuming a uniform shape and constant porosity, can be expressed mathematically in Equations (3.10) and (3.11) [25, 26]. An equilibrium of particle dynamics is reached at the steady-state condition. In this state, a continued particle or microflocs/aggregate attachment to the larger flocs/aggregate is prevented, or the breakup kinetics is equal to the turbulence-induced collision rates.

In assessing aggregate strength and resistance to hydrodynamic-induced breakup, two common approaches are normally followed, namely limiting growth and limiting strength. The former relies on the determination of the maximum floc size before rupture, while the latter is based on the micromechanical analysis of aggregate strength. Many empirical and theoretical formulations based on the mentioned approaches are available in the literature (Equations 3.12 and 3.13). Liu et al. [27] presented the yield stress approach for calculating maximum aggregate tensile strength $\boldsymbol{\tau}_y$ at which breakage is likely to occur in the inertia range of turbulence (Equation 3.12). Similarly, Attia [28] presented modes for predicting the critical fluid velocity above which there will be aggregate disruption by estimating floc yield stress σ_y resulting from the dynamic pressure acting on the floc (Equation 3.13).

$$\sigma = \mu G = \mu\sqrt{\frac{\varepsilon}{v}} \tag{3.10}$$

$$\tau = \frac{(1-p)F_B}{p{d_p}^2} \tag{3.11}$$

$$\tau_y = \frac{F_B}{d_p^{\,2}}\left(\frac{\lambda_0}{d_F}\right)^k \tag{3.12}$$

$$\sigma_y = \frac{1}{2}\rho_f v^2 \tag{3.13}$$

3.5 VORTEX PATTERNS IN FLUID–PARTICLE REACTORS

The flow vortex patterns in many fluid–particle reactors are quite diverse and their analysis and characterisation can provide an insight into the fluid and particle dynamics within the reactor. One important feature in many of these flow types is the presence of rotation or swirling or a combination of both, resulting in anisotropic turbulence. A few of the dominant vortex patterns will be discussed in this section. It is also worth mentioning that the flow pattern is a function of the reactor geometry, stirrer or agitator, baffles, and the operating conditions. The focus will be on the common reactor geometries and flow regimes that are typically encountered in many practical applications.

A qualitative and quantitative analysis of the flow pattern in a processing reactor may be performed experimentally using optical diagnostic imaging techniques such as the Particle Image Velocimetry (PIV) or Laser Doppler Velocimetry (LDV) or by numerical modelling. The selected approach in a given scenario will depend on the required detail of the flow field as well as the available resources. While the latter provide a flexible option for the investigation of fluid flow problems, the former often complements or is used in the validation of a numerical model. A detailed description of specific techniques is beyond the scope of this submission. A few of the vortex patterns in some of these reactors are discussed below.

3.5.1 Rotatory Flow Vortex Pattern (e.g., Tubular Reactors)

This type of flow pattern is commonly found in rotating tubular reactors with an enclosed flow induced by a stirrer. Previous studies have shown that this flow pattern has a profound influence on the performance of fluid–particle reactors, and such a type of flow pattern can also be found in many technical applications [29, 39, 52]. For instance, in terms of the stirrer-vessel configuration, the flocculation performance is significantly influenced by the impeller type and its speed. The axial impeller has been found to promote floc formation over a range of impeller speeds as it produces a more homogeneous distribution of local velocity gradients in comparison to the radial impeller. Also, high velocity gradients occur in the region around the impeller, which might significantly hamper the high-rate activated sludge (HRAS) flocculation process [53, 54].

3.5.2 Oscillatory Flow Vortex Pattern (e.g., Oscillatory Reactors)

An oscillator is a device whose action causes intermittent velocity gradients over time, space, and direction, and can be used for colloid removal based on physical flow manipulation. An oscillatory flow pattern is commonly found in reactors with an enclosed flow induced by an oscillator. Results from previous studies have shown that gentle oscillation can promote simultaneous flocculation and particle agglomeration over a relatively short period of time [42]. This technique, whose implementation can result in reductions to reactor sizes and process times, has a strong potential to improve conventional separation processes. It has also been observed that higher oscillating frequency promotes faster upward vertical velocities,

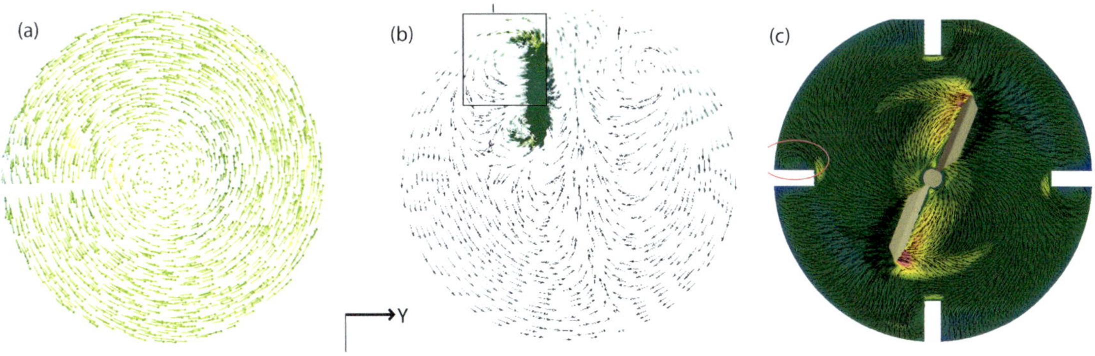

FIGURE 3.6 Typical cross-sections of vortex patterns in fluid–particle reactors: (a) rotatory flow field (reprinted with permission from Springer [29]), (b) oscillatory flow field (reprinted with permission from Taylor & Francis [43]), and (c) swirling and recirculating flow field.

resulting in different sedimentation patterns and removal efficiencies generated by different oscillation frequencies [43].

3.5.3 Swirling Flow Vortex Pattern (e.g., Stirred Tanks, Hydrocyclones)

A swirling type of flow pattern is commonly found in hydrocyclones [55–57] and counter-rotating reactors with an enclosed flow induced by a rotational swirling effect [58, 59]. In addition to hydrocyclones, this type of flow field is also common in rotating reactors with baffles such as mixing tanks, in which case, the flow is a combination of rotational and swirl-dominated flow as shown in Figure 3.6c. In the case of a counter-rotating vortex reactor, it has also been found that the swirl ratio and micromixing time of the flow increase as the Multi-Inlet Vortex Reactor (MIVR) is scaled up, indicating a flow with stronger swirl yet less mixing efficiency [60]. In order to promote mixing and enhance the floc formation process, some baffles should be installed to break water flow co-rotation with the impeller [54]. In modifying the flow pattern to suit a particular condition, it is advisable to perform the optimisation of the impeller shape in parallel with vessel geometry [61, 62].

3.6 CASE STUDIES IN ENVIRONMENTAL PROCESS ENGINEERING

This section presents a short review of the recent studies on the applications of turbulence or hydrodynamics in environmental and biological systems (Table 3.2). Water purification, irrigation, water quality assessment, sludge dewatering, bio-flocculation, and bio clogging are some of the technical areas of application identified. It should be noted that improvements in the performance of the engineered processes (e.g., stirred tanks, shear reactors, and photobioreactors) that rely on the identified areas of applications continue to shape the research focus in the field of environmental process engineering [29]. In this respect, studies have been conducted to determine how to accurately quantify the impact of hydrodynamic characteristics on the infectivity of bacteriophage MS2, a norovirus surrogate. Several studies also involved the development of reactors for testing the effect of hydrodynamic characteristics on microalgae and human enteric viruses, which cannot be easily prepared in large quantities [29–33]. The results obtained from the studies indicated that hydrodynamic cavitation could trigger the inactivation of waterborne

TABLE 3.2 Selected Studies on Computational and Experimental Applications of Vortex Dynamics

TECHNICAL APPLICATION	*FLOW REGIME*	*STUDY TYPE*	*REFERENCES*
Biofouling/biofilms	Turbulent	Experimental/CFD	[25–30]
Water disinfection/irrigation	Turbulent	Experimental/CFD	[31–35]
Water self-purification	Turbulent	Experimental	[43–46]
Solid–liquid separation	Turbulent	Experimental	[2, 36–38]
Food and paper processing	Turbulent	Experimental	[39–42]

viruses to levels defined in water quality directives. This was reportedly due to OH radicals that form an Advanced Oxidation Process (AOP) during the cavitation process and high shear forces inside the cavitation structure. Also, flow structures in a hydrodynamic filter have been numerically investigated [34]. In this study, the tangential component of velocity was defined, and the three-dimensional pattern of the current/streamlines was obtained using their two-dimensional projection in the meridian cross-section of the filter, which allows one to discover the vortex structures. It was concluded that the optimal flow regime can be implemented by selecting the optimal correlation between the flow of liquid regime to be processed and the rotation frequency of the filtering baffle in the hydrodynamic filter. The remaining sub-sections describe how hydrodynamics, turbulence, and vortex dynamics are applied to achieve the desired process efficiency in other identified areas of applications—water purification, sludge dewatering, food processing, and self-purification of the water bodies.

3.6.1 Hydrodynamics in Water Treatment Process

In fluid engineering problems, research has consistently focused on identifying parameters that improve engineered processes including water purification and inactivation of pathogens [35]. While the conventional technique of disinfection by chlorination has been employed to kill pathogenic microorganisms in raw water, recent studies have shown that chlorine reacts with organic compounds in water and generates disinfection by-products (DBPs), such as trihalomethanes (THMs), haloacetic acids (HAAs), and so on. Consequently, turbulence-induced inactivation has been studied as an alternative approach.

The effects of hydrodynamic parameters such as orifice size, orifice number, and orifice layout of multi-orifice plate, cavitation number, cavitation time, and orifice velocity on the microbial population have been investigated to determine how the desired process efficiency can be achieved [36]. Experimental results have shown that cavitation effects increased with a decrease in orifice size and an increase in orifice number, cavitation time, and orifice velocity. Flow hydrodynamics and pipe material have also been shown to influence biofilm development in drinking water distribution systems (DWDS). Furthermore, biofilm development was inhibited at higher flows, indicating that shear forces imposed by the flow conditions were above the critical levels for biofilm attachment. Experimental data from these studies were used to characterise the hydrodynamic behaviour for numerical simulation and validation [37, 38]. A low-cost, pipe-based, pathogen reduction system was also demonstrated by Thomas et al. Their approach has huge potential for application in developing countries due to its simple design [39].

3.6.2 Hydrodynamics in Solid–Liquid Separation

In the design of process reactors, it is often necessary to tailor the separation technique to the dynamics and characteristics of the working slurry. Hence, several studies have been conducted to determine the influence of hydrodynamics on sludge processing. The optimum dosage values, which were obtained when flocculation performance was assessed based on surrogate indicators such as sludge volume index and supernatant turbidity, confirmed polymer bridging as the primary flocculation mechanism. Specific apparatus construction and reactor geometry were found to help maintain sludge suspension in a metastable state that is crucial for the formation of pellet-like compact agglomerates with better dewaterability and uniform aggregate size [2, 40]. Similarly, sludge disintegration, using a rotor-stator type hydrodynamic cavitation reactor (HCR), has been experimentally investigated [41]. To determine the effects of cavitation (including thermal energy) and shear stress on sludge disintegration, the performance of the HCR with and without the dimples and temperature control was analysed. The results indicated that when dimples were present and there was no temperature control, the reduction of sludge particles increased by 50%–80%. Furthermore, the disintegration performance increased with the rotational speed and was minimised at the highest inlet pressure. Several other studies leveraging on the hydrodynamics of the process reactor for fluid–particle separation are available in the literature [42, 43]. Many of these lab-scale studies have demonstrated the feasibility of turbulence-induced fluid–particle separation.

3.6.3 Hydrodynamics in Food and Paper Processing

Hydrodynamic cavitation (HC) is a process in which high energy is released in a flowing liquid upon bubble implosion due to a decrease and subsequent increase in local pressure. In food and beverage industries, hydrodynamic cavitators can be utilised for the purpose of extraction, emulsification, sterilisation, disinfection, and homogenisation [44]. HC, which can effectively induce sonochemistry by mechanical means, creates extraordinarily high pressures of ~1000 bar, local hotspots with ~5000 K, and high oxidation (hydroxyl radicals) in a room environment, without introducing new chemicals. For possible industrial applications, the efficiency of HC has been studied by comparing the chemical oxygen demand (COD) removal efficiency of a Venturi device to that of an orifice plate. A sucrose solution and an effluent from a sucrose-based soft drink industry were treated. Results showed that the Venturi device recorded 90% COD removal efficiency after three treatment minutes. On the other hand, the orifice plate recorded 90% COD removal efficiency after 9 min [44–46]. Developing high-performance HCRs and revealing the corresponding disinfection mechanisms constitute the most crucial issues today [45]. Refining cellulose pulp is a critical step in obtaining high-quality paper characteristics; however, this process is slow and costly, especially for refining longer conifer fibres, which are the preferred source for high-quality paper production and give the paper its strength. Recently, hydrodynamic cavitation has been applied for refining conifer-rich pulp samples [47].

3.6.4 Hydrodynamics in Self-Purification of Water Bodies

The self-purification ability of water bodies is related to the prevailing hydrodynamic conditions. Coupled hydrodynamic and water quality models have been used to investigate the spatial and temporal water quality variations of the water bodies. Using an acoustic Doppler velocimeter, the efficiency of an aeration plug-flow device (APFD) in terms of water flow and dissolved oxygen (DO) has been determined experimentally [48–50]. Recent findings have shown that discharges from several rivers flowing into the New York/New Jersey (NY/NJ) harbour interact and interfere with one another. Such interactions can improve or inhibit water and contaminant flushing from the harbour. In Poyang Lake, three-dimensional velocities at various locations as well as the velocity distribution and turbulence characteristics were assessed, and plug-flow characteristics were analysed. The following two patterns of velocity and turbulence in horizontal sections are observed: (1) near the aeration plug-flow device (APFD), the water flow was intensively pushed downstream and simultaneously recirculated; (2) farther away, the reflux area gradually decreased, and the velocity and turbulence distribution trended towards uniform. At the interfaces between two immiscible fluids—water and alkane of small carbon number, the amphiphilic polyethylene oxide (PEO) chain diffuses laterally, experiencing hydrodynamic drags from both phases. The absolute values of interfacial diffusion coefficients demonstrate a bigger contribution from the hydrodynamics of the water phase, which may be attributed to a stronger attraction between water and the PEO molecules [51].

3.7 CONCLUSIONS AND FUTURE PERSPECTIVE

This brief communication is a summary of the main concepts involved in fluid–particle research and technical applications. It is by no means an exhaustive contribution, and readers interested in the details of the subject matter are advised to consult other scientific information available on the subject matter. Fluid process engineering is a field of active research, and there is an ever-increasing scope for the application of hydrodynamics, turbulence, and vortex dynamics. In addition to the identified areas of application, there are several emerging areas of application. As turbulence research advances with better computing power and algorithms, it is hoped that there will be limitless scope for the application of vortex dynamics in fluid process engineering.

DATA AVAILABILITY

Supporting CFD data for this study is available on request from the author.

REFERENCES

1. C.T. Crowe, T.R. Troutt, J.N. Chung, Particle interactions with vortices, in: S.I. Green (ed.), *Fluid Vortices*, Springer, Dordrecht, 1995: pp. 829–861. https://doi.org/10.1007/978-94-011-0249-0_19.
2. B.A. Oyegbile, *Optimization of Micro Processes in Fine Particle Agglomeration by Pelleting Flocculation*, CRC Press, Leiden, 2016. https://doi.org/10.1201/9781315671871.
3. B. Oyegbile, P. Ay, S. Narra, Flocculation kinetics and hydrodynamic interactions in natural and engineered flow systems: A review, *Environ. Eng. Res.* 21 (2016) 1–14. https://doi.org/10.4491/eer.2015.086.
4. B. Oyegbile, G. Akdogan, M. Karimi, Modelling the dynamics of granular particle interactions in a vortex reactor using a coupled DPM-KTGF model, *S. Afr. J. Chem. Eng.* (2020). https://doi.org/10.1016/j.sajce.2020.05.008.
5. J. Morchain, Bioreactor modeling, *Bioreactor Modeling* (2017) 29–84. https://doi.org/10.1016/B978-1-78548-116-1.50002-1.
6. E.L. Paul, V.A. Atiemo-Obeng, S.M. Kresta (eds.), *Handbook of Industrial Mixing: Science and Practice*, John Wiley & Sons, Hoboken, NJ, 2004. https://doi.org/10.1002/0471451452.
7. Z. Harun, E.R. Lotfy, Generation, evolution, and characterization of turbulence coherent structures, in: R. Barillé (ed.), *Turbulence and Related Phenomena*, InTechOpen, 2019. https://doi.org/10.5772/intechopen.76854.
8. G.H. Yeoh, C.P. Cheung, J. Tu, *Multiphase Flow Analysis Using Population Balance Modeling: Bubbles, Drops and Particles*, Butterworth-Heinemann, Waltham, MA, 2014.
9. S.M. Peker, Ş.Ş. Helvacı, H.B. Yener, B. İkizler, A. Alparslan (eds.), *Solid–Liquid Two Phase Flow*, Elsevier, Oxford, 2008. https://doi.org/10.1016/B978-0-444-52237-5.X5001-2.
10. K. Boodhoo, A. Harvey, Process intensification: An overview of principles and practice, in: *Process Intensification for Green Chemistry: Engineering Solutions for Sustainable Chemical Processing*, John Wiley & Sons, 2013. https://doi.org/10.1002/9781118498521.ch1.
11. M. Zlokarnik, *Stirring: Theory and Practice*, Wiley-VCH, Weinheim, 2008.
12. N. Kockmann, *Transport Phenomena in Micro Process Engineering*, Springer, Heidelberg, 2008.
13. J.Y. Oldshue, R.R. Trussell, R. Trussell, Design of impellers for mixing, in: A. Amirtharajah, M.M. Clark, R. Trussell (eds.), *Mixing in Coagulation and Flocculation*, American Water Works Association, Denver, CO, 1991: pp. 309–342.
14. B.E. Logan, *Environmental Transport Processes*, John Wiley & Sons, Hoboken, NJ, 2012.

15. D.H. Bache, R. Gregory, *Flocs in Water Treatment*, IWA Publishing, London, 2007. https://doi.org/10.2166/9781780402000.
16. J. Bratby, *Coagulation and Flocculation in Water and Wastewater Treatment*, IWA Publishing, London, 2016.
17. G.H. Yeoh, J. Tu, *Computational Techniques for Multiphase Flows*, 2nd ed., Elsevier, Oxford, 2019. https://doi.org/10.1016/C2017-0-01655-5.
18. P. Taboada-Serrano, C.-J. Chin, S. Yiacoumi, C. Tsouris, Modeling aggregation of colloidal particles, *Curr. Opin. Colloid Interface Sci.* 10 (2005) 123–132. https://doi.org/10.1016/j.cocis.2005.07.003.
19. J. Gregory, *Particles in Water: Properties and Processes*, CRC Press, Boca Raton, FL, 2006. https://doi.org/10.1201/9780203508459.
20. H.R. Norouzi, R. Zarghami, R. Sotudeh-Gharebagh, N. Mostoufi, *Coupled CFD–DEM Modeling: Formulation, Implementation and Application to Multiphase Flows*, John Wiley & Sons, Chichester, 2016. https://doi.org/10.1002/9781119005315.
21. C.T. Crowe, J.D. Schwarzkopf, M. Sommerfeld, Y. Tsuji, *Multiphase Flows with Droplets and Particles*, CRC Press, Boca Raton, FL, 2011. https://doi.org/10.1201/b11103.
22. D.N. Thomas, S.J. Judd, N. Fawcett, Flocculation modelling: A review, *Water Res.* 33 (1999) 1579–1592. https://doi.org/10.1016/S0043-1354(98)00392-3.
23. J.S. Marshall, Effect of particle collisions on the expulsion of heavy particles from a vortex core, *Phys. Fluids* 18 (2006) 113301. https://doi.org/10.1063/1.2370427.
24. R. Sun, H. Xiao, H. Sun, Investigating the settling dynamics of cohesive silt particles with particle-resolving simulations, *Adv. Water Resour.* 111 (2018) 406–422. https://doi.org/10.1016/j.advwatres.2017.11.012.
25. C. Coufort, D. Bouyer, A. Liné, Flocculation related to local hydrodynamics in a Taylor–Couette reactor and in a jar, *Chem. Eng. Sci.* 60 (2005) 2179–2192. https://doi.org/10.1016/j.ces.2004.10.038.
26. P.A. Shamlou, N. Titchener-Hooker, Processing of solid–liquid suspensions, *Processing of Solid–Liquid Suspensions* (1993) 1–25. https://doi.org/10.1016/B978-0-7506-1134-3.50005-3.
27. S.X. Liu, L.A. Glasgow, Aggregate disintegration in turbulent jets, *Water Air Soil Pollut.* 95 (1997) 257–275. https://doi.org/10.1007/BF02406169.
28. Y.A. Attia, J.S. Laskowski, J. Ralston, Flocculation, in: *Colloid Chemistry in Mineral Processing*, Elsevier, Amsterdam, 1992: pp. 277–308.
29. Q. Zhang, X. Wu, S. Xue, K. Liang, W. Cong, Study of hydrodynamic characteristics in tubular photobioreactors, *Bioprocess Biosyst. Eng.* 36 (2012) 143–150. https://doi.org/10.1007/s00449-012-0769-2.
30. J. Kosel, I. Gutiérrez-Aguirre, N. Rački, T. Dreo, M. Ravnikar, M. Dular, Efficient inactivation of MS-2 virus in water by hydrodynamic cavitation, *Water Res.* 124 (2017) 465–471. https://doi.org/10.1016/j.watres.2017.07.077.
31. Q.H. Zhang, X. Wu, S.Z. Xue, Z.H. Wang, C.H. Yan, W. Cong, Hydrodynamic characteristics and microalgae cultivation in a novel flat-plate photobioreactor, *Biotechnol. Prog.* 29 (2012) 127–134. https://doi.org/10.1002/btpr.1641.
32. H. Yan, C. Guan, Y. Jia, X. Huang, W. Yang, Mixing characteristics, cell trajectories, pressure loss and shear stress of tubular photobioreactor with inserted self-rotating helical rotors, *J. Chem. Technol. Biotechnol.* 93 (2017) 1261–1269. https://doi.org/10.1002/jctb.5484.
33. A. Muller-Feuga, J. Pruvost, R.L. Guédes, L.L. Déan, P. Legentilhomme, J. Legrand, Swirling flow implementation in a photobioreactor for batch and continuous cultures of *Porphyridium cruentum*, *Biotechnol. Bioeng.* 84 (2003) 544–551. https://doi.org/10.1002/bit.10818.
34. V.A. Devisilov, E.Y. Sharai, Numerical study of the flow structure in a hydrodynamic filter, *Theor. Found. Chem. Eng.* 50 (2016) 209–216. https://doi.org/10.1134/s0040579516020044.
35. M.P. Baranova, Hydrodynamic cavitation unit for improving the performance of irrigation water during plant growing, *IOP Conf. Ser. Earth Environ. Sci.* 315 (2019) 062018. https://doi.org/10.1088/1755-1315/315/6/062018.
36. T. Kudryashova', S. Polyakov, N. Tarasov, A novel parallel algorithm for 3D modelling electromagnetic purification of water, *Matec Web Conf.* 210 (2018) 04027. https://doi.org/10.1051/matecconf/201821004027.
37. Z. Dong, W. Zhao, Killing rate of colony count by hydrodynamic cavitation due to square multi-orifice plates, *IOP Conf. Ser. Earth Environ. Sci.* 121 (2018) 022004. https://doi.org/10.1088/1755-1315/121/2/022004.
38. M.W. Cowle, G. Webster, A.O. Babatunde, B.N. Bockelmann-Evans, A.J. Weightman, Impact of flow hydrodynamics and pipe material properties on biofilm development within drinking water systems, *Environ. Technol.* (2019) 1–13. https://doi.org/10.1080/09593330.2019.1619844.
39. S.F. Thomas, P. Rooks, F. Rudin, N. Cagney, S. Balabani, S. Atkinson, et al., Swirl flow bioreactor containing dendritic copper-containing alginate beads: A potential rapid method for the eradication of *Escherichia coli* from waste water streams, *J. Water Process Eng.* 5 (2015) 6–14. https://doi.org/10.1016/j.jwpe.2014.10.010.

40. B. Oyegbile, P. Ay, S. Narra, Optimization of physicochemical process for pre-treatment of fine suspension by flocculation prior to dewatering, *Desal. Water Treat.* 57 (2016) 2726–2736. https://doi.org/10.1080/19443994.2015.1043591.
41. H. Kim, B. Koo, X. Sun, J.Y. Yoon, Investigation of sludge disintegration using rotor-stator type hydrodynamic cavitation reactor, *Sep. Purif. Technol.* 240 (2020) 116636. https://doi.org/10.1016/j.seppur.2020.116636.
42. E. Halfi, A. Brenner, D. Katoshevski, Separation of colloidal minerals from water by oscillating flows and grouping, *Sep. Purif. Technol.* 210 (2019) 981–987. https://doi.org/10.1016/j.seppur.2018.08.054.
43. E. Halfi, A. Arad, A. Brenner, D. Katoshevski, Development of an oscillation-based technology for the removal of colloidal particles from water: CFD modeling and experiments, *Eng. Appl. Comp. Fluid* 14 (2020) 622–641. https://doi.org/10.1080/19942060.2020.1748114.
44. N. Asaithambi, P. Singha, M. Dwivedi, S.K. Singh, Hydrodynamic cavitation and its application in food and beverage industry: A review, *J. Food Process Eng.* 42 (2019). https://doi.org/10.1111/jfpe.13144.
45. X. Sun, J. Liu, L. Ji, G. Wang, S. Zhao, J.Y. Yoon, et al., A review on hydrodynamic cavitation disinfection: The current state of knowledge, *Sci. Total Environ.* 737 (2020) 139606. https://doi.org/10.1016/j.scitotenv.2020.139606.
46. P.H.L. Alves, P. de S.L. Silva, D.C. Ferreira, J.C. de S.I. Gonçalves, COD removal from sucrose solution using hydrodynamic cavitation and hydrogen peroxide: A comparison between Venturi device and orifice plate, *Rev. Bras. Recursos Hídricos* 24 (2019). https://doi.org/10.1590/2318-0331.241920180147.
47. J. Kosel, A. Šinkovec, M. Dular, A novel rotation generator of hydrodynamic cavitation for the fibrillation of long conifer fibers in paper production, *Ultrason. Sonochem.* 59 (2019) 104721. https://doi.org/10.1016/j.ultsonch.2019.104721.
48. B. Li, G. Yang, R. Wan, H. Li, Hydrodynamic and water quality modeling of a large floodplain lake (Poyang Lake) in China, *Environ. Sci. Pollut. Res.* 25 (2018) 35084–35098. https://doi.org/10.1007/s11356-018-3387-y.
49. Y. Li, H. Feng, H. Zhang, J. Sun, D. Yuan, L. Guo, et al., Hydrodynamics and water circulation in the New York/New Jersey Harbor: A study from the perspective of water age, *J. Mar. Syst.* 199 (2019) 103219. https://doi.org/10.1016/j.jmarsys.2019.103219.
50. X. Li, M. Huang, R. Wang, Numerical simulation of Donghu Lake hydrodynamics and water quality based on remote sensing and MIKE 21, *ISPRS Int. J. Geo-Inf.* 9 (2020) 94. https://doi.org/10.3390/ijgi9020094.
51. Z. Li, J. Yang, J.V. Hollingsworth, J. Zhao, Lateral diffusion of single polymer molecules at interfaces between water and oil, *RSC Adv.* 10 (2020) 16565–16569. https://doi.org/10.1039/d0ra02630a.
52. F. Visscher, J. van der Schaaf, T.A. Nijhuis, J.C. Schouten, Rotating reactors – A review, *Chem. Eng. Res. Des.* 91 (2013) 1923–1940. https://doi.org/10.1016/j.cherd.2013.07.021.
53. J. Bridgeman, B. Jefferson, S.A. Parsons, Computational fluid dynamics modelling of flocculation in water treatment: A review, *Eng. Appl. Comp. Fluid* 3 (2014) 220–241. https://doi.org/10.1080/19942060.2009.11015267.
54. S. Balemans, S.E. Vlaeminck, E. Torfs, L. Hartog, L. Zaharova, U. Rehman, et al., The impact of local hydrodynamics on high-rate activated sludge flocculation in laboratory and full-scale reactors, *Processes* 8 (2020) 131. https://doi.org/10.3390/pr8020131.
55. S.M. Mousavian, A.F. Najafi, Influence of geometry on separation efficiency in a hydrocyclone, *Arch. Appl. Mech.* 79 (2008) 1033–1050. https://doi.org/10.1007/s00419-008-0268-8.
56. G.Q. Dai, J.M. Li, W.M. Chen, Numerical prediction of the liquid flow within a hydrocyclone, *Chem. Eng. J.* 74 (1999) 217–223. https://doi.org/10.1016/S1385-8947(99)00044-3.
57. Y. Liu, Q. Cheng, B. Zhang, F. Tian, Three-phase hydrocyclone separator – A review, *Chem. Eng. Res. Des.* 100 (2015) 554–560. https://doi.org/10.1016/j.cherd.2015.04.026.
58. J. Derksen, Confined and agitated swirling flows with applications in chemical engineering, *Flow Turbul. Combust.* 69 (2002) 3–33. https://doi.org/10.1023/A:1022419316418.
59. S. Jakirlic, K. Hanjalic, C. Tropea, Modeling rotating and swirling turbulent flows: A perpetual challenge, *AIAA J.* 40 (2002) 1984–1996. https://doi.org/10.2514/2.1560.
60. Z. Liu, M. Ramezani, R.O. Fox, J.C. Hill, M.G. Olsen, Flow characteristics in a scaled-up multi-inlet vortex nanoprecipitation reactor, *Ind. Eng. Chem. Res.* 54 (2015) 4512–4525. https://doi.org/10.1021/ie5041836.
61. M. Jaszczur, A. Młynarczykowska, L. Demurtas, Effect of impeller design on power characteristics and Newtonian fluids mixing efficiency in a mechanically agitated vessel at low Reynolds numbers, *Energies* 13 (2020) 640. https://doi.org/10.3390/en13030640.
62. I.A. Escamilla-Ruíz, F.Z. Sierra-Espinosa, J.C. García, A. Valera-Medina, F. Carrillo, Experimental data and numerical predictions of a single-phase flow in a batch square stirred tank reactor with a rotating cylinder agitator, *Heat Mass Transfer* 53 (2017) 2933–2949. https://doi.org/10.1007/s00231-017-2030-7.

Numerical Modelling and Physicochemical Characterisation of Fluid–Particle Systems

4

Benjamin Oyegbile

4.1 INTRODUCTION

Hydrodynamic and physicochemical interactions play an important role in many industrial unit processes and hence their importance in many engineering applications of fluid flow. Fluid flow investigations in a wide range of process conditions as well as complex biological, physical, and chemical processes have been the subject of many scientific publications over the past two decades. Several studies on bench, pilot, and industrial scales have been conducted on a wide variety of hydrodynamic conditions and different reactor geometric designs. In many of these studies, the aim is to provide an insight into the fluid flow and process dynamics in terms of the spatial and temporal evolution within the flow device, and in some cases, performance testing of newly designed flow units and processing techniques with potential applications on an industrial scale. Regardless of the focus of these studies, it is quite apparent that valuable information can be obtained from the basic study of fluid flow dynamics in process units, especially from design and optimisation perspectives.

A quick survey of the study in this field shows that many innovative process reactors have been successfully tested on different scales for a wide variety of technical applications ranging from fine particle separation and water purification to cell culture preparation [1–6]. Experimental data, which are collected in these studies for numerical validation purposes, are often used to characterise the hydrodynamic behaviour as well as to quantify the fluid parameters of interest, such as the flow velocity profile, vorticity, turbulent kinetic energy and its rate of dissipation, turbulent intensity, and so on. While there is a large body of scientific literature focusing on the hydrodynamics and physicochemical processes in stirred tank reactors, the aim of the present communication is to briefly summarise developments in this field, especially in the application of the knowledge of the fluid dynamics to fluid–particle reactor design, development, and optimisation.

DOI: 10.1201/9781003518426-4

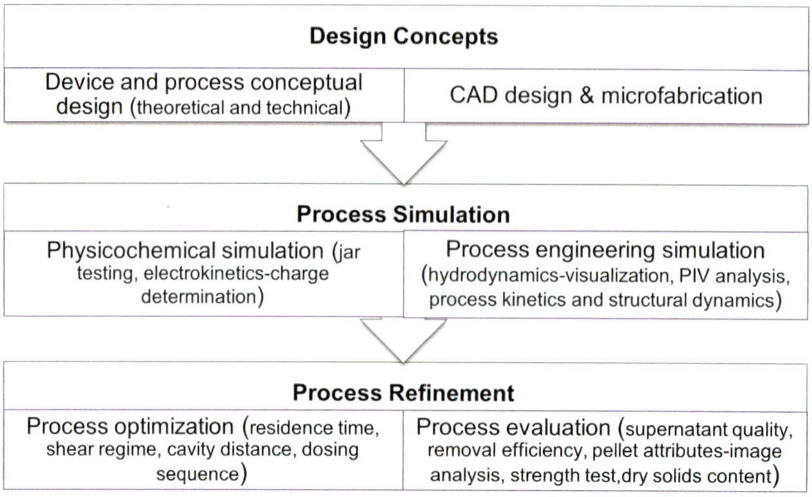

FIGURE 4.1 Reactor design and process optimisation workflow for fluid–particle systems.

4.2 DESIGN, MODELLING, AND OPTIMISATION OF FLUID–PARTICLE REACTORS

4.2.1 Reactor Design Considerations and Process Optimisation

In fluid engineering problems, research has shown that it is possible to optimise all influencing process parameters in an evolutionary manner, right from the conceptual design to the final performance testing phase. This will entail the integration of the fluid flow investigation into the process reactor conceptual design and system optimisation [1]. Nowadays, this multistage process design and optimisation workflow shown in Figure 4.1 can be fully automated through the use of a computational platform. In formulating and developing a numerical solution strategy to a particular physical problem involving fluid–particle interactions, a sound theoretical understanding and analysis of the problem are often required. This will assist in the selection of appropriate experimental data collection methods and mathematical models that sufficiently encapsulate the physics of the problem. A number of numerical approaches and solution strategies discussed in the subsequent sections have been developed for a multitude of fluid flow scenarios. Therefore, it is important to evaluate each circumstance individually and form an opinion regarding which model would provide the best fit for a particular fluid engineering problem. It has been suggested that the robustness of any mathematical model is a function of the numerical code being used and the flow scenario being modelled [7].

4.2.2 Fluid Dynamics and Governing Equations

The interactions of different phases in fluid flow occur on different scales of the fluid motion, as depicted in Figure 4.2. Fluid dynamics is primarily concerned with the macroscopic phenomena of fluid flow in which the fluid is treated as a continuum. For instance, a fluid element is composed of many molecules, and the fluid dynamics represent the behaviour of the numerous molecules within the system. This concept, with certain assumptions, forms the basis of the derivation of fluid conservation equations of mass

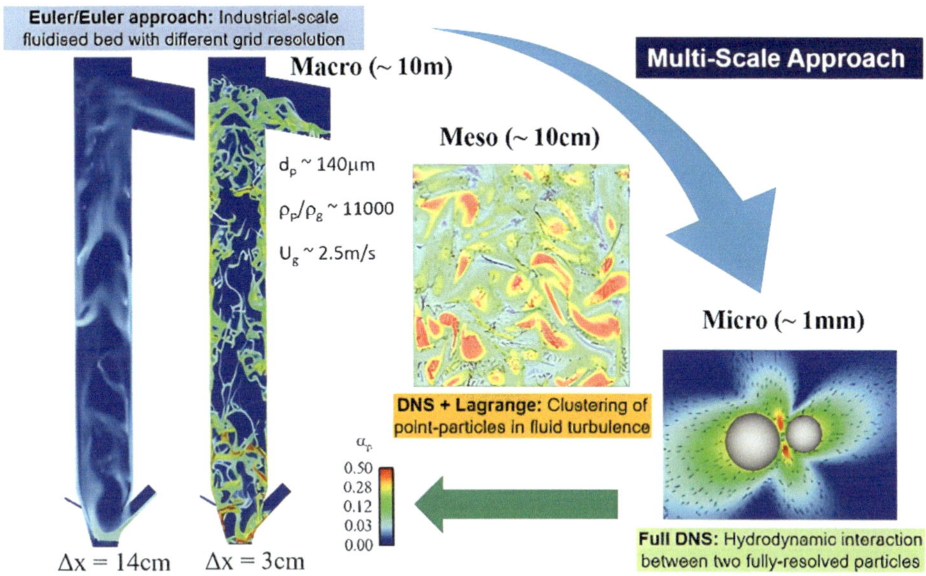

FIGURE 4.2 Multiscale modelling approach in the simulation of fluid–particle system. Reproduced with permission from Springer [14].

and momentum, also known as the Navier–Stokes equation, using a fluid control volume [8, 9]. The general form of the Reynolds-averaged Navier–Stokes (RANS) equation of mass and momentum conservation in any fluid flow system can be written as follows:

$$\frac{\partial \overline{v}_i}{\partial x_i} = 0 \tag{4.1}$$

$$\rho\frac{\partial \overline{v}_i}{\partial t} + \rho\overline{v}\frac{\partial \overline{v}_i}{\partial x_j} = -\frac{\partial \overline{p}}{\partial x_i} + \frac{\partial}{\partial x_j}(\overline{\tau}_{ij} - \rho\overline{v_i'v_j'}) + \rho\vec{g} + F \tag{4.2}$$

In Equations (4.1) and (4.2), ρ is the density, v is the mean velocity, $\overline{p}$ is the mean pressure, $\overline{\tau}_{ij}$ is the viscous stress tensor, $\rho\overline{v_i'v_j'}$ is the Reynolds stresses, which represent the transfer of momentum due to turbulent fluctuations, t is the time, $\vec{g}$ is the gravitational body force, while F_σ represents any external body force acting on the fluid.

4.2.3 Numerical Simulation and Solution Strategies

In modelling complex multiphase flows in mixing tanks and process reactors, there exist two common numerical solution strategies, namely Eulerian–Eulerian and Eulerian–Lagrangian formulations, depending on the scale of the fluid flow being investigated as shown schematically in Figure 4.2. In the former, the fluid domain is treated as an interpenetrating continuum, while in the latter, the discrete or distinct particles of the dispersed phase are tracked in the Lagrangian reference frame. In addition to the flow field information, the particle population information such as the mean size, mass or volume fraction, and number density can be obtained using either the population balance or the discrete phase models [10, 11].

Several variants of particle tracking exist in the Eulerian–Lagrangian class, such as the Eulerian granular model based on the kinetic theory of granular flow (KTGF), the disperse phase model (DPM), the discrete element model (DEM), and the macroscopic particle model (MPM). In the case of the Eulerian–Eulerian approach, the species distribution of the discrete phase may be accounted for using the population balance model (PBM), while the Eulerian–Lagrangian models can directly compute the particle size distribution while taking into account different collision and interaction mechanisms using DEM [15–18].

4.2.4 Treatment of Flow Domain and Turbulence

Turbulence modelling forms an integral part of the numerical analysis of complex fluid flows since most engineering fluid flows entail a certain form of instability. Several closure models have been developed for resolving turbulence parameters in steady-state Reynolds-averaged Navier–Stokes (RANS) equations. The two-equation RANS models such as $k - \varepsilon$ and $k - \omega$ have been found to perform reasonably well in the modelling of rotating flows in many process reactors, with the only drawback being the assumption of local isotropic turbulence. The underlying theoretical assumptions underpinning the use of these models can be found in the following reference materials [12, 13]. Since the reactors encountered in most practical physicochemical processes contain moving or rotating parts, it is therefore necessary to take this into consideration in the preparation of the computational grid. The most common strategies for steady-state calculations include the single reference frame (SRF), multiple reference frame (MRF) or frozen rotor approach, mixing plane model , and snapshot approach, while the sliding or dynamic mesh is frequently used in transient calculations of fluid flow. For detailed information on the practical applications of the above-mentioned methods, readers are referred to the following reference texts [20, 21]. Figure 4.3 provides a summary of the different scales of fluid motion and turbulence encountered in many engineering flows.

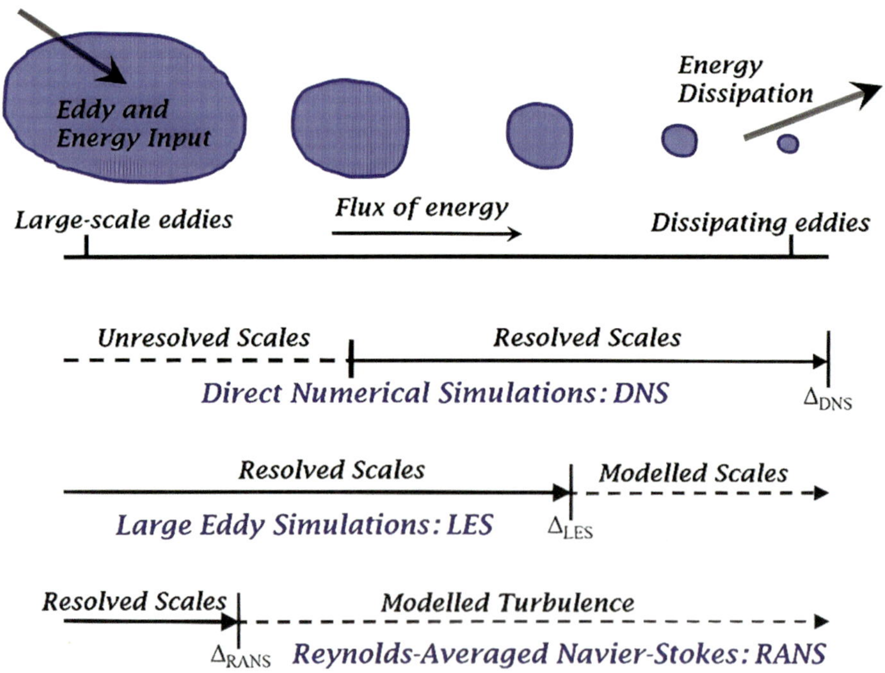

FIGURE 4.3 Fluid scale and turbulence resolution in multiphase modelling. Reproduced with permission from CRC Press [14].

4.2.5 Model Coupling in Multiphase Flow Problems

Modelling complex physicochemical processes involving fluid flow sometimes necessitates the integration of the existing mathematical models in order to appropriately describe the physics of the problem. This can be achieved through the use of specially developed or customised in-house numerical codes or a modification of the existing ones, with several software package vendors offering a platform for software improvement through the use of Application Programming Interface (API) or Application Customisation Toolkit (ACT). Such flexibility allows engineers and researchers to extend the capability and versatility of the existing numerical codes. Many software vendors go a step further in this respect by actively encouraging the development of scalable apps that extend the capability of their core software; an excellent example is the mixing tank template released by ANSYS Inc. for the automation of mixing tank simulation processes. However, there exist several other flexible options for numerical code development using the open-source platform such as OpenFOAM, and the readers are advised to consider available options for their specific problems.

4.3 EXPERIMENTAL QUANTIFICATION OF PHYSICOCHEMICAL PARAMETERS

Several analytical and instrumental techniques have been developed for the study of complex hydrodynamic-mediated processes that often occur across a wide range of process conditions in flocculation, wet agglomeration, sedimentation, floatation, fluidisation, and crystallisation. The techniques shown in Figure 4.4 are used either in the quantification of the hydrodynamics of the carrier and dispersed phase or in the determination of the spatial and temporal evolution of the discrete phase properties, such as the change in the particle size and distribution. In the case of the hydrodynamic interactions of the carrier and dispersed phase, a number of laser-based fluid flow techniques such as particle image velocimetry (PIV), particle tracking velocimetry (PTV), laser Doppler anemometry (LDA), laser Doppler velocimetry (LDV), and, more recently, radioactive tracking techniques such as positron emission particle tracking (PEPT) and computer-aided radioactive particle tracking (CARPT) have gained wider acceptance in the scientific and industrial communities due to their ease of use and non-intrusive nature [20–22]. These

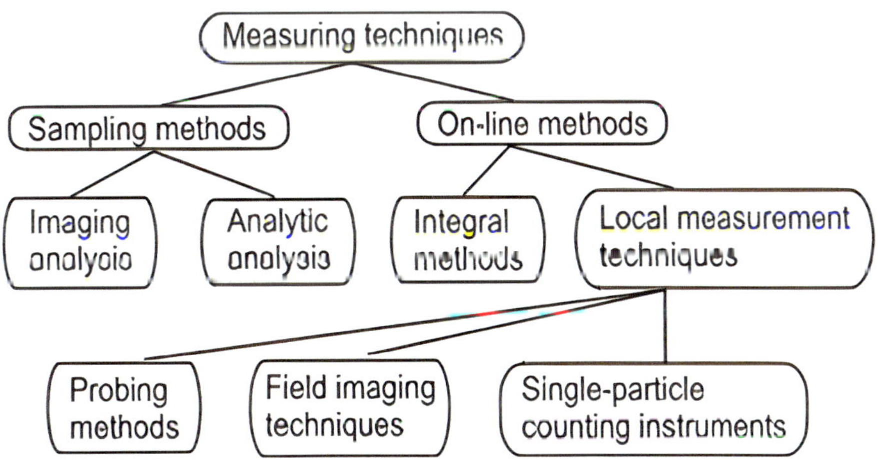

FIGURE 4.4 Experimental measurement techniques for particle-laden flow. Reprinted with permission from Taylor & Francis [18].

techniques provide valuable insight into the salient macroscale fluid flow characteristics, such as the instantaneous and time-averaged hydrodynamic behaviour of the continuous phase, as well as the influence of the dispersed phase on the fluid flow. This is achieved by coupling the flow field measurements with the particulate phase properties and motion [21]. The experimental data set is subsequently used in the validation of numerical simulation results [18, 23].

The dominant and widely used macroscale experimental fluid flow characterisation techniques are laser velocimetry and radioactive particle tracking techniques such as the PIV or PTV, LDV, or LDA, with the PIV reported to be a more efficient technique [24]. These online methods facilitate the determination of the properties of multiphase particle-laden flow, especially at low concentration. These local methods are quite superior to other similar techniques, such as optical fibre probing and light scattering, due to their non-intrusive nature with little or no interference on the flow while providing time series and time-averaged fluid flow characteristics with a high spatial resolution [18]. The workings of a typical field imaging technique, such as PIV, consist of the tracer particles, laser source for flow illumination, and high-capacity cameras—complementary metal-oxide semiconductor (CMOS) or charge-coupled device (CCD) for the fluid flow image recording. The captured images are thereafter post-processed and correlated to obtain the hydrodynamic parameters of interest. Table 4.1 provides a list of recent studies on the experimental analysis of physicochemical processes in stirred tanks. These studies demonstrated the importance of robust and reliable experimental data for complex fluid flow analysis and numerical model validation. Recent advances in experimental techniques have led to the emergence of radioactive particle tracing measurement techniques, which aim to improve the ease of data collection, data accuracy, and reliability.

In order to correlate the hydrodynamic and process conditions with the suspension or dispersion properties, especially the change in the species concentration—spatial and temporal evolution of the

TABLE 4.1 Selected Studies on the Experimental Analysis of Physicochemical Processes

REACTOR CONFIGURATION	*STIRRER CONFIGURATION*	*EXPERIMENTAL TECHNIQUE*	*TECHNICAL APPLICATION*	*TRACER PARTICLES*
Cylindrical tank	Rotating disc	2D PIV	Mixing/ agglomeration	Silver-coated and hollow glass spheres [1]
Cylindrical tank	Hydrofoil impellers	PEPT	Mixing	Radioactive particles [26]
Cylindrical tank	Rushton turbine	PIV	Mixing	Polymeric and glass particles [27]
Cylindrical tank	Hollow blade semi-elliptic disc turbine	TRPIV, PIV	Mixing	Neutrally buoyant glass beads [24]
Cylindrical tank	Pitched-blade turbine impeller	FPIV	Mixing	Soda-lime glass beads [28]
Cylindrical tank	Rushton turbine	CARPT	Mixing	Radioactive particles [29]
Cylindrical tank	Rushton turbine	LDA	Mixing	Hollow glass spheres [30]
Cylindrical tank	Kenics static mixer	PEPT	Mixing	Radioactive particles [31]
Cylindrical tank	Pitched-blade turbine	PEPT	Mixing	Radioactive particles [32]
Cylindrical tank	Pitched-blade turbine	PIV	Mixing	Silica glass spheres [23]
Square tank	Hydro foil impeller	PIV, image analysis	Mixing/ agglomeration	In situ agglomerated flocs [33]
Cylindrical tank	Six-blade Rushton turbine	3V3	Mixing	Opt image polycrystalline particles [34]
Cylindrical tank	Rushton turbine, pitched-blade turbine	PEPT	Mixing	Monosized silica gel particles [35]
Cylindrical tank	Six-blade Rushton turbine	PEPT, LDA	Mixing	Ion-exchange resin particles [36]
Cylindrical tank	Rotor-stator mixer	PIV	Mixing	Polyamide particles [37]

particle size distribution, a number of laboratory measurement techniques are widely adopted [25]. The choice will depend to a large extent on the concentration and size distribution of the disperse phase and the nature of the flow. Regardless of the chosen analytical approach, such a correlation will facilitate an assessment of the treatment process and the reactor performance under a particular process condition. For instance, the conventional physicochemical simulation tests such as the cylinder, Imhoff cone and jar tests can be combined with parametric analytical techniques such as the Buchner-funnel or pressure filtration test, capillary suction time (CST) test, electrokinetic charge analysis using colloidal titrations (i.e., zeta and streaming potential), laser light scattering or laser diffraction, microscopy, image analysis, photometric dispersion analysis (PDA), fibre optic sensor, and Proton Nuclear Magnetic Resonance (^{1}H NMR) spectroscopy. These techniques have been successfully employed to characterise the physicochemical process in bench, pilot, and full-scale studies [38–41]. A careful consideration of the limitations of each of these approaches will ensure the proper selection of an appropriate method.

In most of the physicochemical processes involving particulate flow either as a colloidal dispersion or as a granular suspension, the species attributes—mean size, particle concentration and distribution, and fractal properties of the resulting agglomerates—are the primary parameters of interest [21]. In this case, an appropriate physicochemical simulation such as a jar or cylinder test is often followed by a parametric analysis to characterise the process performance as a function of species attributes. Several other parameters may be of interest depending on the type of reactor and the required solid–liquid separation method. Such parameters may include aggregate mean size, shape and distribution, aggregate volume concentration, aggregate strength, sludge volume index, silting index, residual supernatant turbidity, absorbance or optical density, electrical conductivity, viscosity, zeta or streaming potential, specific resistance to filtration, capillary suction time, and so on [38, 39]. In the case of chemical optimisation, a parametric dose–response curve will give reasonably accurate information on the required chemical dose for a particular process condition [42–45]. Table 4.2 and Figure 4.5 show a typical correlation of the agglomerate characteristic properties such as the particle size distribution (PSD) with the process condition—shear rate. However, regardless of the choice of parametric test, an examination of the supernatant, sediment, filtrate, and residue will yield some valuable information on the reactor performance under specific process conditions. Such assessment is carried out either by direct *in situ* measurements such as in particle counting, *ex situ* analysis in which the samples are extracted for measurements, or by other indirect parametric indicators. A more detailed discussion on the practical applications of different dispersed phase measurement techniques is available elsewhere [40, 41].

Considering the wide range of options available to select from, optimising a given physicochemical condition for a particular process reactor under laboratory conditions is a daunting task. Therefore, in optimising the design and process parameters for a particular reactor, a statistical correlation of these parameters from a data set is often required, depending on the available time and complexity of the problem, to obtain accurate information on the optimum design and process conditions. A number of statistical methods such as the design of experiments and response surface methodology can be applied to a large set of experimental data to obtain the desired optimisation points. This will facilitate an understanding of the influence of different process conditions on the reactor performance, which will assist in the selection of optimised process conditions.

TABLE 4.2 Aggregate Properties as a Function of Reactor's Operating Conditions

	AGITATION SPEED	
TEST PARAMETERS	*145 RPM*	*165 RPM*
Mean agglomerate diameter (mm)	3.8330	3.9182
Mean agglomerate compressive strength (Nm m^{-2})	0.4298	0.4351
Mean strain rate (s^{-1})	0.3639	0.4088
Mean maximum compressive force (N)	4.9476	5.2303

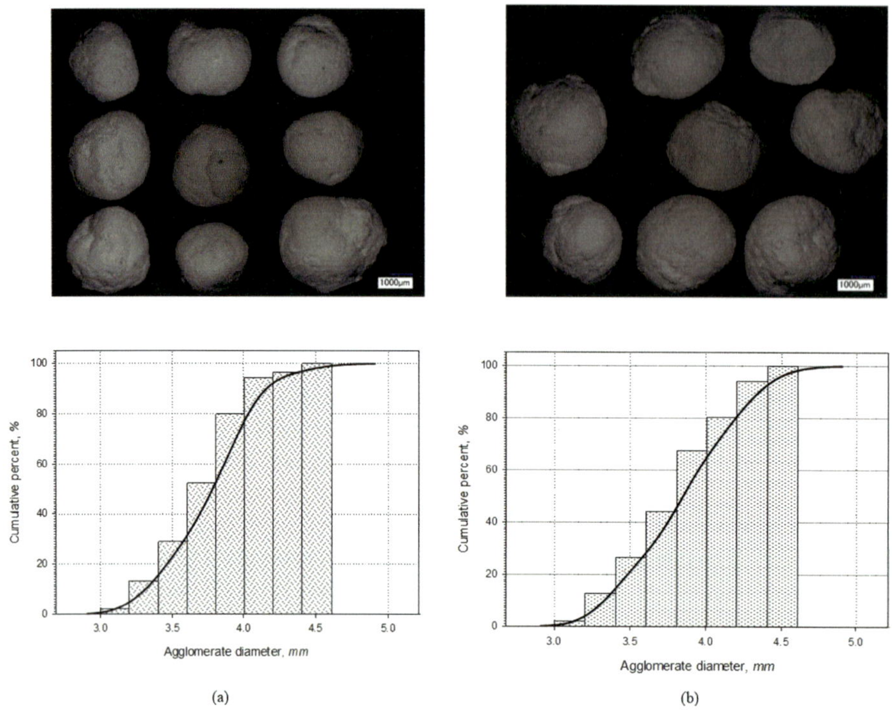

FIGURE 4.5 Parametric correlation of aggregate properties and the process conditions: (a) 145 rpm and (b) 165 rpm.

4.4 MODELLING PHYSICOCHEMICAL PROCESSES IN STIRRED TANK REACTORS

4.4.1 CFD Modelling of Physicochemical Process

The use of computational fluid dynamics (CFD) as a research tool to investigate complex fluid–particle interactions in many industrial processes has been growing in popularity in both academia and the industry [46]. CFD provides a powerful alternative and a more robust platform for engineers in the design of equipment and processes involving fluid flow and heat transfer when compared to the classical experimental approach. Nowadays, numerical simulations complement the experimental measurements and are increasingly being performed in many fluid engineering applications ranging from chemical and mineral processing to civil and environmental process engineering [46]. However, it is worth pointing out that the continual development of reliable empirical, mathematical, and computational models relies on robust and detailed experimental data.

Table 4.3 provides a list of experimentally validated numerical simulation studies focusing on the physicochemical analysis of fluid–particle reactors, which are focused on the technical application of mixing for several industrial processes. The modelling approach in most of these studies is applicable to mixing tanks and other process reactors of various geometric designs. Joshi et al. [47, 48] provide a comprehensive review of CFD applications in single-phase mixing tank hydrodynamic analysis, focusing on axial and radial flow impellers in a multitude of flow scenarios. Their two-part study, which is one of the most detailed and comprehensive reviews in this field, summarises developments in mixing tank

TABLE 4.3 Selected Studies on the Numerical Simulation of Physicochemical Processes

REACTOR TYPE/ CONFIGURATION	*STIRRER TYPE/APPLICATION*	*EXPERIMENTAL VALIDATION METHOD*	*NUMERICAL CODE/ MODELLING APPROACH*	*TURBULENCE MODELS*
Cylindrical flocculator	Paddle mixer/flocculation	LDA, 2D PIV	CFX/MRF	$k-\varepsilon$, RSM [53]
Rectangular flocculator	Axial impeller/water purification	2D PIV	Fluent/MRF	$k-\varepsilon$ [54]
Cylindrical sedimentation tank	Axial impeller/water purification	Laser Diffraction	CFX/MRF	$k-\varepsilon$ [55]
Cylindrical jar testing device	Paddle stirrer/flocculation	LDA	Fluent/MRF	$k-\varepsilon$, $k-\omega$, RSM [7, 56, 57]
Cylindrical flocculation reactor	Rushton turbine/bio-flocculation	LDV	Fluent/MRF	$k-\varepsilon$ [58]
Cylindrical stirred tank	Pitched turbine blade/silica particle deagglomeration	Laser diffraction/PIDS	Fluent/MRF	$k-\varepsilon$[59]
Cylindrical stirred bioreactor	Marine impeller/cell cultivation	Tracer and dynamic method	Fluent/MRF	$k-\varepsilon$ [60]
Cylindrical tank	R1342-type impeller/flocculation	Image analysis	Fluent/MRF	$k-\varepsilon$ [61]
Cylindrical tank	Rushton impeller/cell culture	Optical sensor	CFX/MRF	$k-\varepsilon$ [62]
Cylindrical bioreactor	Rushton, scaba, and paddle impellers/cell culture	Optical density	Fluent/MRF	$k-\varepsilon$ [63]
Cylindrical tank	Turbine, anchor, and oblique impellers/autoclave	Tracer injection	Fluent/MRF	$k-\varepsilon$ [64]
Cylindrical bioreactor	Marine impeller/recombinant protein synthesis	PIV	Fluent/MRF	$k-\varepsilon$ [60, 65]
Cylindrical tube reactor	Impeller/bacterial inactivation	2D PIV	CFX/MRF	RSM [2]
Cylindrical bioreactor	Rushton turbine/anaerobic digestion	Gas chromatography	Fluent/MRF	$k-\varepsilon$ [66]
Cylindrical tank	Turbine impeller/polymerisation	Droplet size measurements	Fluent/MRF	$k-\varepsilon$ [67]
Cylindrical tank	Rushton impeller/cell cultivation	Dynamic method	Fluent/MRF	$k-\varepsilon$ [68]
Cylindrical tank	Rushton turbine/cell inactivation	PIV	Fluent/MRF	$k-\varepsilon$ [69]
Cylindrical tank	Rushton turbine and propellers/cell culture	Dynamic method	Fluent/MRF	RSM [70]
Cylindrical crystalliser	Rushton impeller/precipitation	X-ray/laser diffraction	Fluent/MRF	$k-\varepsilon$ [71]
Cylindrical photobioreactor	Rotating cylinder/algal culture	Optical density	Fluent/SRF	$k-\omega$ [72, 73]

modelling by bringing together in a two-part publication the results of scientific investigations spanning several decades. Similar reviews focusing on turbulent multiphase flows and multiphase reactor modelling that provide a more comprehensive discussion on the subject matter are available elsewhere [19, 49–51].

Regardless of the specific focus of each study, most of the scientific studies in this field differ only in terms of stirrer-vessel configurations, experimental validation methods, and the choice of modelling approach. In terms of the stirrer-vessel configuration, there is a wide variety of flow inducers available for fluid flow investigation, each with different power demands and flow patterns. In addition to well-established impeller designs employed in most of the studies—Rushton turbine, pitched-blade turbine, propeller, etc., a few innovative designs have been used with good results [52]. The turbulence models of choice in most of the investigations are the two-equation RANS models such as $k - \varepsilon$ and $k - \omega$ due to their robustness and lower computational demand as well as the Response Surface Methodology (RSM) model, which is quite efficient in handling rotating flows in stirred tanks and process reactors. The dominant modelling approaches for problems involving moving parts and rotation are the MRF and sliding mesh. The former is suitable for steady-state problems, while the latter is usually employed for transient calculations. Despite the technical limitations of some of the experimental flow measurement techniques, reasonable agreement was obtained in most of the studies between the experimental data and numerical simulation. In a few of the studies, the model predictions were not robust enough when compared to the experimental data set partly due to the complexity of the flow scenario being modelled.

4.4.2 Data-Driven Process Modelling

The classical data-driven modelling, as the name implies, has its roots in experimental measurements and data analysis. Data collection and analysis are commonly used in process design and improvements as they provide insight into the process performance. The rapid development of many statistical software packages, coupled with the emergence of machine learning and artificial intelligence, has rapidly transformed the field of data science. Nowadays, data-driven models have the capability to understand the underlying patterns and behaviour of a system without prior knowledge when adequate and high-quality data is available. This has provided a very powerful platform for predictive modelling of many processes by observing the underlying patterns in the data set. Sophisticated machine learning algorithms such as recurrent neural networks (RNNs) or physics-informed neural networks (PINNs) have the potential to understand the physics behind the data by learning the mechanistic relationships from first principles. This will ultimately improve their predictive capacity and their capability to be used autonomously for process optimisation and control in the future.

4.5 CONCLUSIONS AND FUTURE PERSPECTIVES

This chapter discussed the main concepts involved in fluid–particle design, modelling, and optimisation with respect to industrial applications. It is by no means an exhaustive contribution, and readers interested in the details of the subject matter are advised to consult other scientific information available on the subject matter. The recent developments in this field clearly show that significant progress has been made over the past decade in the understanding of complex physicochemical phenomena that are vital for many industrial and environmental processes, especially from experimental and theoretical perspectives. Fluid process engineering is a field of active research, and there is an ever-increasing scope for the application of hydrodynamics, turbulence, and vortex dynamics in process engineering. In addition to the identified areas of application, there are several emerging areas of application.

However, there is still a gap in knowledge, especially in the suitability of the existing mathematical models to accurately predict the reactor performance in a wide range of existing and emerging industrial processes. This clearly calls for numerical code programming and development to form an integral part of the engineering training and curriculum in the future. The successful design, development, and optimisation of industrial process reactors depend on the robustness of experimental data, mathematical models, and simulation tools. As turbulence research advances with more accessible computing power and algorithms, it is hoped that there will be limitless scope for the application of vortex dynamics in fluid process engineering with the potential to revolutionise many industrial processes. In conclusion, numerical simulation along with robust experimental data and advanced machine learning algorithms will continue to be highly indispensable well into the foreseeable future.

DATA AVAILABILITY

Data sharing is not applicable to this chapter as no new data were created or analysed in this study.

REFERENCES

1. B. Oyegbile, M. Hoff, M. Adonadaga, B. Oyegbile, Experimental analysis of the hydrodynamics, flow pattern and wet agglomeration in rotor-stator vortex separators, *J. Environ. Chem. Eng.* 5 (2017) 2115–2127. https://doi.org/10.1016/j.jece.2017.04.016.
2. S.F. Thomas, P. Rooks, F. Rudin, N. Cagney, S. Balabani, S. Atkinson, et al., Swirl flow bioreactor containing dendritic copper-containing alginate beads: A potential rapid method for the eradication of Escherichia coli from waste water streams, *J. Water Process Eng.* 5 (2015) 6–14. https://doi.org/10.1016/j.jwpe.2014.10.010.
3. M. Sievers, S.M. Stoll, C. Schroeder, M. Niedermeiser, T.I. Onyeche, Sludge dewatering and aggregate formation effects through Taylor vortex assisted flocculation, *Sep. Sci. Technol.* 43 (2008) 1595–1609. https://doi.org/10.1080/01496390801973888.
4. X. Wang, P. Jin, H. Yuan, E. Wang, N. Tambo, Pilot study of a fluidized-pellet-bed technique for simultaneous solid/liquid separation and sludge thickening in a sewage treatment plant, *Water Sci. Technol.* 49 (2004) 81–88.
5. D.D. Dionysiou, G. Balasubramanian, M.T. Suidan, A.P. Khodadoust, I. Baudin, J.-M. Laîné, Rotating disk photocatalytic reactor: Development, characterization, and evaluation for the destruction of organic pollutants in water, *Water Res.* 34 (2000) 2927–2940. https://doi.org/10.1016/S0043-1354(00)00022-1.
6. G. Loraine, G. Chahine, C.T. Hsiao, J.-K. Choi, P. Aley, Disinfection of gram-negative and gram-positive bacteria using DynaJets® hydrodynamic cavitating jets, *Ultrason. Sonochem.* 19 (2012) 710–717. https://doi.org/10.1016/j.ultsonch.2011.10.011.
7. J. Bridgeman, B. Jefferson, S.A. Parsons, Computational fluid dynamics modelling of flocculation in water treatment: A review, *Eng. Appl. Comp. Fluid* 3 (2009) 220–241. https://doi.org/10.1080/19942060.2009.11015267.
8. T. Krüger, H. Kusumaatmaja, A. Kuzmin, O. Shardt, G. Silva, E.M. Viggen, *The Lattice Boltzmann Method: Principles and Practice*, Springer, Basel, 2017.
9. L. Jahanshaloo, E. Pouryazdanpanah, N.A.C. Sidik, A review on the application of the lattice Boltzmann method for turbulent flow simulation, *Numer. Heat Transf. Part A Appl.* 64 (2013) 938–953. https://doi.org/10.1080/10407782.2013.807690.
10. G. Lian, S. Moore, L. Heeney, Population balance and computational fluid dynamics modelling of ice crystallisation in a scraped surface freezer, *Chem. Eng. Sci.* 61 (2006) 7819–7826. https://doi.org/10.1016/j.ces.2006.08.075.
11. S. Das, H. Bai, C. Wu, J.-H. Kao, B. Barney, M. Kidd, et al., Improving the performance of industrial clarifiers using three-dimensional computational fluid dynamics, *Eng. Appl. Comp. Fluid* 10 (2016) 130–144. https://doi.org/10.1080/19942060.2015.1121518.
12. ANSYS, Inc., *ANSYS Fluent Theory Guide*, 18.2, ANSYS, Inc., 2017.
13. ANSYS, Inc., *ANSYS Fluent User's Guide*, 18.2, ANSYS, Inc., 2017.

14. M. Sommerfeld, Numerical methods for dispersed multiphase flows, in: T. Bodnár, G.P. Galdi, Š. Nečasová (eds.), *Part. Flows*, Springer, Heidelberg; 2017: pp. 327–396.
15. R.I. Jeldres, P.D. Fawell, B.J. Florio, Population balance modelling to describe the particle aggregation process: A review, *Powder Technol.* 326 (2017) 190–207. https://doi.org/10.1016/j.powtec.2017.12.033.
16. A. Hellesto, M. Ghaffari, B. Balakin, A parametric study of cohesive particle agglomeration in a shear flow—Numerical simulations by the discrete element method, *J. Dispersion Sci. Technol.* 38 (2016) 611–620. https://doi.org/10.1080/01932691.2016.1185015.
17. D. Schellander, *CFD Simulations of Particle Laden Flow: Particle Transport and Separation*, Anchor Academic Publishing, Hamburg, 2014.
18. C.T. Crowe, J.D. Schwarzkopf, M. Sommerfeld, Y. Tsuji, *Multiphase Flows with Droplets and Particles*, 2nd ed., CRC Press, Boca Raton, FL, 2011.
19. J.B. Joshi, K. Nandakumar, Computational modeling of multiphase reactors, *Annu. Rev. Chem. Biomol. Eng.* 6 (2015) 347–378. https://doi.org/10.1146/annu rev-chembioeng-061114-123229.
20. R.W. Johnson (ed.), *Handbook of Fluid Dynamics*, 2nd ed., CRC Press, Boca Raton, FL, 2016.
21. E. Michaelides, C.T. Crowe, J.D. Schwarzkopf (eds.), *Multiphase Flow Handbook*, 2nd ed., CRC Press, Boca Raton, FL, 2016.
22. P. Mavros, Flow visualization in stirred vessels: A review of experimental techniques, *Chem. Eng. Res. Des.* 79 (2001) 113–127. https://doi.org/10.1205/02638760151095926.
23. G. Li, Z. Gao, Z. Li, J. Wang, J.J. Derksen, Particle-resolved PIV experiments of solid-liquid mixing in a turbulent stirred tank, *AIChE J.* 64 (2018) 389–402. https://doi.org/10.1002/aic.15924.
24. X. Liu, Y. Bao, Z. Li, Z. Gao, Analysis of turbulence structure in the stirred tank with a deep hollow blade disc turbine by time-resolved PIV, *Chin. J. Chem. Eng.* 18 (2010) 588–599. https://doi.org/10.1016/S1004-9541(10)60262-5.
25. B.O. Hasan, Breakage of drops and bubbles in a stirred tank: A review of experimental studies, *Chin. J. Chem. Eng.* 25 (2017) 698–711. https://doi.org/10.1016/j.cjche.2017.03.008.
26. Y.S. Fangary, M. Barigou, J.P.K. Seville, D.J. Parker, A Langrangian study of solids suspension in a stirred vessel by positron emission particle tracking PEPT, *Chem. Eng. Technol.* 25 (2002) 521–528. https://doi.org/10.1002/1521-4125(200205)25:5<521::AID-CEAT521>3.0.CO;2-C.
27. G. Montante, A. Paglianti, F. Magelli, Analysis of dilute solid-liquid suspensions in turbulent stirred tanks, *Chem. Eng. Res. Des.* 90 (2012) 1448–1456. https://doi.org/10.1016/j.cherd.2012.01.009.
28. H. Unadkat, C.D. Rielly, G.K. Hargrave, Z.K. Nagy, Application of fluorescent PIV and digital image analysis to measure turbulence properties of solid-liquid stirred suspensions, *Chem. Eng. Res. Des.* 87 (2009) 573–586. https://doi.org/10.1016/j.cherd.2008.11.011.
29. A. Rammohan, A. Kemoun, M. Al-Dahhan, M. Dudukovic, Characterization of single phase flows in stirred tanks via computer automated radioactive particle tracking (CARPT), *Chem. Eng. Res. Des.* 79 (2001) 831–844. https://doi.org/10.1205/02638760 152721343.
30. A.E. Komrakova, Z. Liu, M.B. Machado, S.M. Kresta, Development of a zone flow model for the confined impeller stirred tank (CIST) based on mean velocity and turbulence measurements, *Chem. Eng. Res. Des.* 125 (2017) 511–522. https://doi.org/10.1016/j.cherd.2017.07.025.
31. M. Rafiee, M.J.H. Simmons, A. Ingram, H.E. Stitt, Development of positron emission particle tracking for studying laminar mixing in Kenics static mixer, *Chem. Eng. Res. Des.* 91 (2013) 2106–2113. https://doi.org/10.1016/j.cherd.2013.05.022.
32. A. Guida, A.W. Nienow, M. Barigou, PEPT measurements of solid–liquid flow field and spatial phase distribution in concentrated monodisperse stirred suspensions, *Chem. Eng. Sci.* 65 (2010) 1905–1914. https://doi.org/10.1016/j.ces.2009.11.005.
33. J. Kilander, S. Blomström, A. Rasmuson, Spatial and temporal evolution of floc size distribution in a stirred square tank investigated using PIV and image analysis, *Chem. Eng. Sci.* 61 (2006) 7651–7667. https://doi.org/10.1016/j.ces.2006.09.001.
34. K.V. Sharp, D. Hill, D. Troolin, G. Walters, W. Lai, Volumetric three-component velocimetry measurements of the turbulent flow around a Rushton turbine, *Exp. Fluids* 48 (2009) 167–183. https://doi.org/10.1007/s00348-009-0711-9.
35. R. Fishwick, M. Winterbottom, D. Parker, X. Fan, H. Stitt, The use of positron emission particle tracking in the study of multiphase stirred tank reactor hydrodynamics, *Can. J. Chem. Eng.* 83 (2005) 97–103. https://doi.org/10.1002/cjce.5450830117.
36. F. Chiti, S. Bakalis, W. Bujalski, M. Barigou, A. Eaglesham, A.W. Nienow, Using positron emission particle tracking (PEPT) to study the turbulent flow in a baffled vessel agitated by a Rushton turbine: Improving data treatment and validation, *Chem. Eng. Res. Des.* 89 (2011) 1947–1960. https://doi.org/10.1016/j.cherd.2011.01.015.

37. H.H. Mortensen, F. Innings, A. Håkansson, The effect of stator design on flowrate and velocity fields in a rotor-stator mixer—An experimental investigation, *Chem. Eng. Res. Des.* 121 (2017) 245–254. https://doi.org/10.1016/j.cherd.2017.03.016.
38. D.H. Bache, R. Gregory, *Flocs in Water Treatment*, IWA Publishing, London, 2007.
39. J. Bratby, *Coagulation and Flocculation in Water and Wastewater Treatment*, 3rd ed., IWA Publishing, London, 2016.
40. J. Gregory, Flocculation measurement techniques, in: T. Tadros (ed.), *Encyclopedia of Colloid and Interface Science*, Springer, Heidelberg, 2013: pp. 492–525.
41. J. Gregory, Monitoring particle aggregation processes, *Adv. Colloid Interface Sci.* 147–148 (2009) 109–123. https://doi.org/10.1016/j.cis.2008.09.003.
42. D.H. Bache, Y.Q. Zhao, Optimising polymer use in alum sludge conditioning: An ad hoc test, *J. Water Supply Res. Technol.—AQUA* 50 (2001) 29–38.
43. B. Oyegbile, P. Ay, S. Narra, Optimization of physicochemical process for pre-treatment of fine suspension by flocculation prior to dewatering, *Desal. Water Treat.* 57 (2016) 2726–2736. https://doi.org/10.1080/19443994.2015.1043591.
44. F.A. Al Momani, B. Örmeci, Optimization of polymer dose based on residual polymer concentration in dewatering supernatant, *Water Air Soil Pollut.* 225 (2014) 1–11. https://doi.org/10.1007/s11270-014-2154-z.
45. S.R. Ramphal, M.S. Sibiya, Optimization of coagulation-flocculation parameters using a photometric dispersion analyser, *Drinking Water Eng. Sci.* 7 (2014) 73–82. https://doi.org/10.5194/dwes-7-73-2014.
46. J. Tu, G.H. Yeoh, C. Liu, *Computational Fluid Dynamics: A Practical Approach*, Butterworth-Heinemann, Oxford, 2013.
47. J.B. Joshi, N.K. Nere, C.V. Rane, B.N. Murthy, C.S. Mathpati, A.W. Patwardhan, et al., CFD simulation of stirred tanks: Comparison of turbulence models. Part I: Radial flow impellers, *Can. J. Chem. Eng.* 89 (2011) 23–82. https://doi.org/10.1002/cjce.20446.
48. J.B. Joshi, N.K. Nere, C.V. Rane, B.N. Murthy, C.S. Mathpati, A.W. Patwardhan, et al., CFD simulation of stirred tanks: Comparison of turbulence models (Part II: Axial flow impellers, multiple impellers and multiphase dispersions), *Can. J. Chem. Eng.* 89 (2011) 754–816. https://doi.org/10.1002/cjce.20465.
49. S. Balachandar, J.K. Eaton, Turbulent dispersed multiphase flow, *Annu. Rev. Fluid Mech.* 42 (2009) 111–133. https://doi.org/10.1146/annurev.fluid.010908.165243.
50. H. Tinoco, H. Lindqvist, W. Frid, Numerical simulation of industrial flows, in: L. Angermann (ed.), *Numerical Simulation—Exemplary Applications in Computational Fluid Dynamics*, InTech, Rijeka, 2001.
51. M. Pradip, Computational fluid dynamics analysis of turbulent flow, in: I.V. Minin, O.V. Minin (eds.), *Computational Fluid Dynamics: Technology and Applications*, InTech, Rijeka, 2011: pp. 255–292.
52. V. Buwa, A. Dewan, A.F. Nassar, F. Durst, Fluid dynamics and mixing of single-phase flow in a stirred vessel with a grid disc impeller: Experimental and numerical investigations, *Chem. Eng. Sci.* 61 (2006) 2815–2822. https://doi.org/10.1016/j.ces.2005.10.066.
53. J. Korpijärvi, E. Laine, H. Ahlstedt, Using CFD in the study of mixing in coagulation and flocculation, in: H.H. Hahn, E. Hoffmann, H. Odegaard (eds.), *Chemical Water and Wastewater Treatment VI*, Springer, Heidelberg, 2000: pp. 89–99.
54. K. Essemiani, C. De Traversay, Optimisation of the flocculation process using computational fluid dynamics, in: H. Hahn, E. Hoffman, H. Odegaard (eds.), *Chemical Water and Wastewater Treatment VII*, IWA Publishing, London, 2002: pp. 41–49.
55. K. Samaras, A. Zouboulis, T. Karapantsios, M. Kostoglou, A CFD-based simulation study of a large scale flocculation tank for potable water treatment, *Chem. Eng. J.* 162 (2010) 208–216. https://doi.org/10.1016/j.cej.2010.05.032.
56. J. Bridgeman, B. Jefferson, S. Parsons, Assessing floc strength using CFD to improve organics removal, *Chem. Eng. Res. Des.* 86 (2008) 941–950. https://doi.org/10.1016/j.cherd.2008.02.007.
57. J. Bridgeman, B. Jefferson, S.A. Parsons, *The Development and Use of CFD Models for Water Treatment Processes*, St. Julians, Malta, 2007.
58. Z. Yang, Z. Wu, G. Zeng, J. Huang, H. Xu, J. Feng, et al., Assessing the effect of flow fields on flocculation of kaolin suspension using microbial flocculant GA1, *RSC Adv.* 4 (2014) 40464–40473. https://doi.org/10.1039/C4RA04101A.
59. G.N. Özcan-Taşkın, G.A. Padron, D. Kubicki, Comparative performance of in-line rotor-stators for deagglomeration processes, *Chem. Eng. Sci.* 156 (2016) 186–196. https://doi.org/10.1016/j.ces.2016.09.023.
60. S.C. Kaiser, R. Eibl, D. Eibl, Engineering characteristics of a single-use stirred bioreactor at bench-scale: the Mobius CellReady 3L bioreactor as a case study, *Eng. Life Sci.* 11 (2011) 359–368. https://doi.org/10.1002/elsc.201000171.

61. W. He, L. Xue, B. Gorczyca, J. Nan, Z. Shi, Experimental and CFD studies of floc growth dependence on baffle width in square stirred-tank reactors for flocculation, *Sep. Purif. Technol.* 190 (2018) 228–242. https://doi.org/10.1016/j.seppur.2017.08.063.
62. H. Zhang, K. Zhang, S. Fan, CFD simulation coupled with population balance equations for aerated stirred bioreactors, *Eng. Life Sci.* 9 (2009) 421–430. https://doi.org/10.1002/elsc.200800074.
63. H. Azargoshasb, S.M. Mousavi, O. Jamialahmadi, M.S.B. SAS, Experiments and a three-phase computational fluid dynamics (CFD) simulation coupled with population balance equations of a stirred tank bioreactor for high cell density cultivation, *Can. J. Chem. Eng.* 94 (2016) 20–32. https://doi.org/10.1002/cjce.22352.
64. H. Zheng, Z. Huang, Z. Liao, J. Wang, Y. Yang, Y. Wang, Computational fluid dynamics simulations and experimental validation of macromixing and flow characteristics in low-density polyethylene autoclave reactors, *Ind. Eng. Chem. Res.* 53 (2014) 14865–14875. https://doi.org/10.1021/ie502551c.
65. A.O.O. Odeleye, D.T.J. Marsh, M.D. Osborne, G.J. Lye, M. Micheletti, On the fluid dynamics of a laboratory scale single-use stirred bioreactor, *Chem. Eng. Sci.* 111 (2014) 299–312. https://doi.org/10.1016/j.ces.2014.02.032.
66. H. Azargoshasb, S. Mousavi, T. Amani, A. Jafari, M. Nosrati, Three-phase CFD simulation coupled with population balance equations of anaerobic syntrophic acidogenesis and methanogenesis reactions in a continuous stirred bioreactor, *J. Ind. Eng. Chem.* 27 (2015) 207–217. https://doi.org/10.1016/j.jiec.2014.12.037.
67. L. Xie, Q. Liu, Z.-H. Luo, A multiscale CFD-PBM coupled model for the kinetics and liquid–liquid dispersion behavior in a suspension polymerization stirred tank, *Chem. Eng. Res. Des.* 130 (2018) 1–17. https://doi.org/10.1016/j.cherd.2017.11.045.
68. T.K. Villiger, B. Neunstoecklin, D. Karst, E. Lucas, M. Stettler, H. Broly, et al., Experimental and CFD physical characterization of animal cell bioreactors: From micro- to production scale, *Biochem. Eng. J.* (2017). https://doi.org/10.1016/j.bej.2017.12.004.
69. Y. Liu, Z.-J. Wang, J. Xia, C. Haringa, Y. Liu, J. Chu, et al., Application of Euler-Lagrange CFD for quantitatively evaluating the effect of shear force on *Carthamus tinctorius* L. cell in a stirred tank bioreactor, *Biochem. Eng. J.* 114 (2016) 209–217. https://doi.org/10.1016/j.bej.2016.07.006.
70. J. Sarkar, L.K. Shekhawat, V. Loomba, M.S. RAS, CFD of mixing of multi-phase flow in a bioreactor using population balance model, *Biotechnol. Prog.* 32 (2016) 613–628. https://doi.org/10.1002/btpr.2242.
71. U. Ojaniemi, J. Puranen, M. Manninen, E. Gorshkova, M. Louhi-Kultanen, Hydrodynamics and kinetics in semi-batch stirred tank precipitation of l-glutamic acid based on pH shift with mineral acids, *Chem. Eng. Sci.* 178 (2017) 167–172. https://doi.org/10.1016/j.ces.2017.12.029.
72. X. Gao, B. Kong, R.D. Vigil, Comprehensive computational model for combining fluid hydrodynamics, light transport and biomass growth in a Taylor vortex algal photobioreactor: Lagrangian approach, *Bioresour. Technol.* 224 (2017) 523–530. https://doi.org/10.1016/j.biortech.2016.10.080.
73. X. Gao, B. Kong, M. Ramezani, M.G. Olsen, R.D. Vigil, An adaptive model for gas–liquid mass transfer in a Taylor vortex reactor, *Int. J. Heat Mass Transf.* 91 (2015) 433–445. https://doi.org/10.1016/j.ijheatmasstransfer.2015.07.125.

5 Numerical Simulation and Design Optimisation of Hydrodynamics and Mixing Process in a Stirred Tank Reactor (STR)

Benjamin Oyegbile

5.1 INTRODUCTION TO FLUID MIXING

Efficient fluid–particle mixing and dispersion is highly indispensable in many environmental processes such as water and wastewater treatment and sludge treatment. A substantial proportion of process engineering applications relies on the mixing and multiphase flow in stirred tanks. In many of these mixing applications, there is always a need for cost savings arising from power consumption which always necessitates a trade-off between energy consumption, pumping capacity, and degree of mixing or mixing performance. Therefore, energy-efficient mixing will ultimately entail using an impeller that can convect and disperse the fluid constituents uniformly within the fluid with minimal power inputs [1]. In other words, using an impeller that is capable of producing the fastest mixing time with the smallest power consumption. Mixing promotes particle–particle and fluid–particle contact, thereby speeding up mass and momentum transfer [2]. Mixing can be described as a random distribution of two or more separate phases through one another via the transport mechanisms of advection, turbulence, and diffusion occurring at different scales, as illustrated schematically in Figure 5.1. Mixing is typically performed using a variety of methods ranging from simple dispersion, blending, agitation, or sparging to static flow manipulation. Advective fluid transport mechanisms refer to the movement of fluid mass where the inertia of the flow

DOI: 10.1201/9781003518426-5

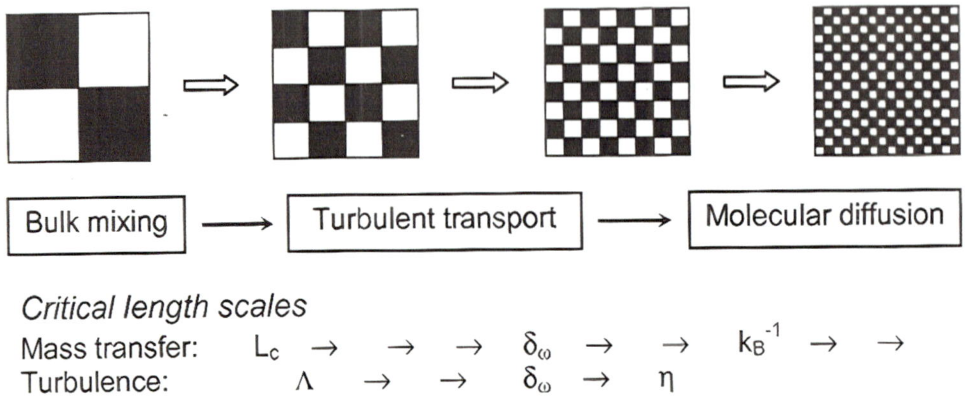

FIGURE 5.1 Schematic illustration of the transport mechanisms in fluid mixing. Reproduced with permission from IWA Publishing [4].

results in large eddy formation. Turbulent fluid mixing on the other hand generates velocity gradients or fluid shear, which results in the formation of eddies and flow vortex. Molecular diffusion is ultimately the only process capable of mixing fluid components on the molecular scale [3].

Fluid flow is generally classified into two distinct classes, namely laminar flow and turbulent flow with the main difference being the value of the dimensionless Reynolds number that describes the state of the flow. Turbulent mixing in stirred vessels typically characterised by high Reynolds number ($Re > 10^4$) can be viewed as a hierarchy of irregular, rotational, and dissipative motion containing vorticity (curl or rotation of the velocity vector) on different scales or eddy sizes. In most turbulent flows such as those encountered in pipe or tube flow, channel flow, and stirred tanks or mixing vessels, energy transfer occurs on different eddy scales [5, 6]. Eddies are spatial recognisable flow patterns that exist in a turbulent flow for at least a short time in which there is a correlation between the velocities at two different points [7]. Fluid deformation causes vortices to stretch and vorticity and kinetic energy to be transported from larger to smaller eddies [8, 9]. The turbulent vortex in such fluid motion is generally propagated in the tangential and axial direction rather than in the radial direction, with the vorticity increasing with decreasing eddy size [5, 9, 10].

In most practical applications of mixing, there exist three scales of mixing, namely macromixing, mesomixing, and micromixing. Macromixing refers to mixing that is driven by the largest scale of motion in the fluid Λ (integral length scale), mesomixing on the other hand involves mixing on a scale smaller than the bulk circulation but larger than the micro mixing, while micromixing is the mixing on the smallest scale of fluid motion λ_0 (Kolmogorov microscale) expressed in Equation (5.1) and at the final scales of molecular diffusivity (Batchelor scale) [11–15]. Micromixing describes the process homogenisation of liquid balls with their surroundings on a molecular level. In typical mixing conditions, the dividing line between micro- and macroscale is between 100 and 1000 μm, respectively [16]. In order to determine the mass transport and transfer rates in the sheared fluid, the kinetic energy dissipation rate ε in the system must be determined from the energy input into the system [17]. Fluid mixing and the subsequent reaction of the fluid constituents are achieved by external power input into the system. The energy-carrying eddies of similar dimension as the agitator diameter ($\Lambda \sim d_A$) introduced into the system by the flow mixer will consequently be dissipated into heat by the smallest eddies of Kolmogorov length scale λ_0 due to the viscous shear. The power input into the system can be calculated using the torque metre measurement from the rotating shaft.

$$\lambda_0 = \left[\frac{\nu^3}{\varepsilon}\right]^{\frac{1}{4}} \tag{5.1}$$

The largest eddies in turbulent dispersion that represent the macroscale of turbulence or the integral length scale (Λ) contain most of the energy and are produced by the stirrer or agitator head. The size of this macroscale of turbulence is of the order of the diameter of the agitator [12, 17, 18]. The turbulent flow can be viewed as an eddy continuum, with its size ranging from the dimension of the turbulence-generating device to the Kolmogorov length scale [18]. In between the energy-containing eddies of the integral length scale Λ at the upper end of the inertia subrange and the smallest eddies of Kolmogorov microscale λ_0 at the beginning of the viscous dissipation range [19], there exist many eddies of other scales smaller than the integral scale Λ that transfer kinetic energy continually through the other length scales. The Batchelor λ_B and Taylor λ_T scales expressed in Equations (5.2) and (5.3) are examples of other important length scales, where D is the diameter of the flocs while ε is the turbulent kinetic energy dissipation rate. The Taylor scale λ_T is an intermediate length scale in the viscous subrange that is representative of the energy transfer from large to small scales, but it is not a dissipation scale and does not represent any distinct group of eddies [20]. Batchelor microscale, on the other hand, is a limiting length scale where the rate of molecular diffusion is equal to the rate of dissipation of turbulent kinetic energy and represents the size of the region within which a molecule moves due to diffusional forces [11, 14, 19, 21]. Figure 5.2 shows different eddy sizes that exist in a turbulent flow and their energy spectrum:

$$\lambda_B = \left[\frac{d^2 \nu}{\varepsilon}\right]^{\frac{1}{4}} \tag{5.2}$$

$$\lambda_T = \sqrt{15}\,\frac{\nu}{\varepsilon}\left(v'\right)_{rms} \tag{5.3}$$

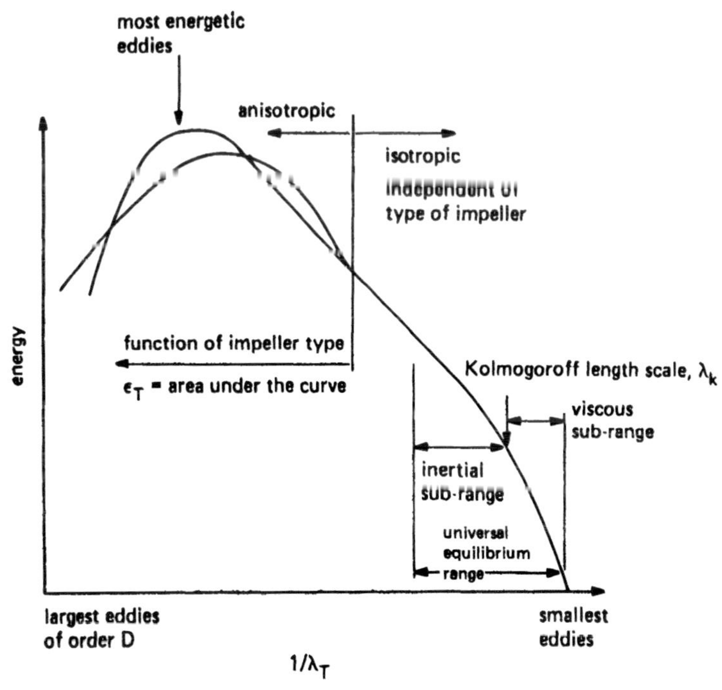

FIGURE 5.2 Eddy sizes and energy spectrum in a turbulent flow. Reprinted with permission from Elsevier B.V. [22].

5.2 HYDRODYNAMIC CHARACTERISATION OF MIXING IN STIRRED TANK

Stirred tanks are widely utilised in chemical, pharmaceutical, and biochemical industries due to their ability to facilitate effective mixing, as well as mass and heat transfer. The performance of a stirred tank reactor is intrinsically linked to the degree of mixing achieved within the vessel. Therefore, optimising the mixing performance of stirred tanks is crucial for achieving desired product quality, maximising product yield, reducing operational costs, and enhancing process efficiency [23]. Several operating parameters such as the agitation speed, impeller torque, impeller tip velocity, power consumption, fluid properties, type of stirrer, and shape of the vessel (stirrer-vessel configuration) have been shown to influence mixing performance in a stirred tank. Consequently, mixing performance depends on complex interactions of these parameters. Theoretical and empirical relationships between these parameters have been developed that allow for rapid assessment of the effect of the operating conditions on the mixing efficiency. For instance, empirical correlations such as power number, Reynolds number or pumping number have been successfully employed in the characterisation and selection of suitable stirrer depending on the mixing and process requirements [24]. Hydrodynamic characterisation of mixing tank impellers, which evolved from the theoretical work on pumps, is based on angular and axial momentum balances around the impeller. The angular momentum balance around the impeller results in torque on the impeller and serves as a measure of power consumption and finally power (N_p) and flow numbers (N_Q). Axial momentum on the other hand determines the axial thrust on the shaft due to the acceleration of the fluid mass from which the momentum number M_o can be derived [25]. In the case of axial flow impellers, these quantities are expressed mathematically in Equations (5.4)–(5.6). Figures 5.3 and 5.4 show an empirical correlation of power and flow numbers with Reynolds number, which can be found useful in the design of new stirrer for a mixing tank:

$$N_p = \frac{P}{\rho N^3 D^5} \tag{5.4}$$

$$N_Q = \frac{Q}{ND^3} \tag{5.5}$$

$$M_0 = 1.43 N_Q^{\,2} \tag{5.6}$$

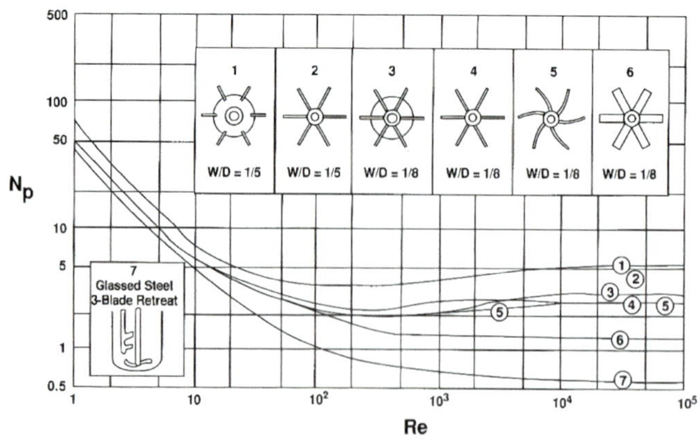

FIGURE 5.3 Power number as a function of the Reynolds number for common impellers. Reprinted with permission from John Wiley [26].

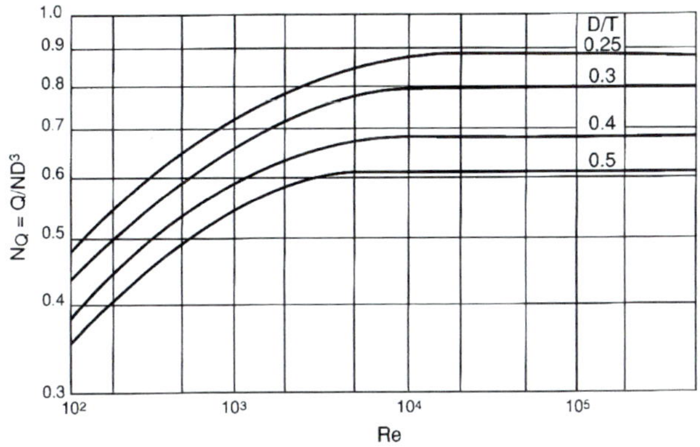

FIGURE 5.4 Flow number as a function of the Reynolds number for pitched-blade turbine. Reprinted with permission from John Wiley [26].

5.3 EMPIRICAL MODELS OF HYDRODYNAMICS AND MIXING TIME

One of the most common indicators of mixing performance in a stirred tank is the concept of mixing time. Mixing efficiency in a stirred tank is generally quantified in terms of the degree of homogeneity, which is defined as the time taken to reach uniform distribution or a specific level of homogeneity (usually taken as 90–95% uniformity) after the release of tracer or perturbation within the fluid being stirred. Several empirical models have been put forward to estimate mixing time and ultimately mixing performance of stirred tank reactors. Common mixing time estimation models include empirical correlations, tracer-dispersion model, circulation-time models, and eddy-scale model [27]. The empirical correlation model expresses mixing time using correlations between different hydrodynamic characteristics of the impeller such as those discussed in the previous paragraph, while the circulation-time models relate mixing time to the bulk circulation of the fluid. Eddy-scale models invoke the turbulence time scalers while the tracer-dispersion models treat mixing as a diffusive–convective transport of tracers. Each of these approaches has its own assumptions, which must be carefully considered when applying them to different mixing scenarios.

Much of the early work on the characterisation of mixing in stirred tanks was focused on the correlation between the mixing time and the flow regime (laminar, transitional, and turbulent flow). The most common empirical correlation of fluid dynamics and mixing time was proposed by Grenville [28]. The mixing time–power number correlation (Equation 5.7) was developed for both transitional and turbulent flow regimes using a mixing tank with single impeller and an aspect ratio of 1. The equation showed, as previously discussed, that at 95% homogeneity, the mixing time depends only on the mixing tank geometry and the power number in the turbulent regime.

$$N_{\theta 95} = 5.2\frac{T^{1.5}H^{1.5}}{D^2 N_p^{\frac{1}{3}}} \quad (5.7)$$

Similarly, Cooke et al. [28] developed a mixing time–turbulence correlation for a multi-impeller stirred tank with an aspect ratio greater than 1 (Equation 5.8). The mixing time obtained by Cooke's correlation was larger than that of Grenville, and this difference was attributed to the development of zone mixing when using multiple radial impellers. The Cooke's correlations also confirm that the mixing time is a function of tank geometry and power number:

$$N_{\theta 95} = \frac{3.3}{N_p^{\frac{1}{3}} N} \left(\frac{H}{D} \right)^{2.43} \tag{5.8}$$

5.4 PREVIOUS STUDIES ON STIRRED TANK HYDRODYNAMICS AND MIXING TIME

Many experimental studies have been conducted on the measurement of blending process and mixing time in stirred tanks using some of the techniques discussed in the preceding sections. Early studies on the measurements of mixing performance were largely based on the experimental measurement of mixing time using a variety of flow tracing techniques, as previously discussed. Kramer et al. [29] in one of the early experimental work on the concept of mixing time showed that the dimensionless mixing time, which is a product of the measured mixing time and the impeller speed, was constant for different impeller speeds in a turbulent flow condition. His work also showed that the dimensionless flow time is independent of the Reynolds number in both laminar and turbulent flow regions [22]. Ghotli et al. [30] developed a predictive model of mixing time by performing measurements of mixing time under turbulent conditions and fitting a mechanistic equation of mixing time to the experimental data to obtain the model constants. The determination of the mixing time was based on the measurement of the concentration of tracer in the mixing tank. A fuzzy C-means predictive model was developed to predict the mixing time and study the influence of the operating parameters such as the impeller design and off-bottom clearance on the mixing time.

With the advances in the development of computational hardware and computational fluid dynamics (CFD) software codes, mixing analysis and optimisation are increasingly being performed using the CFD platform [31]. This has led to the proliferation of CFD software codes and the emergence of many powerful open-source codes such as OpenFOAM and its variants, which have helped CFD to gain wider acceptance beyond academia. Buwa et al. [1] carried out experimental measurement of mixing time and power consumption as well as computational modelling of the flow field and mixing performance of the grid disc impeller. Upon successful validation, the model was subsequently used to investigate the effects of the impeller rotational speed and the grid disc configuration on the fluid dynamics and mixing performance of the grid disc impeller. The predicted flow properties such as the mean velocities, power consumption, and mixing time predicted by the computational model showed good agreement with the experimental measurements. The results of the computational experiment showed that the modifications of the grid disc impeller configuration (radial vs square spaced) do not have any significant effect on the power consumption and mixing efficiency.

In a similar investigation, Yang et al. [32] employed a novel six-blade grid disc impeller (RT-G) and a standard Rushton impeller (RT) to study single-phase turbulent flow and mixing in a stirred tank using Detached Eddy Simulation (DES). They compared the fluid dynamic characteristics of the two impellers in terms of the mean flow velocities, turbulent kinetic energy, power consumption, and mixing time. The CFD model showed a good agreement with published experimental data using the same set of impellers. The results of their study showed an increase of 20–30% in the turbulent kinetic energy and a decrease of 11% in the mixing time of the grid disc impeller when compared to the Rushton impeller at the same level of power consumption, which clearly showed that the grid disc impeller showed a superior energy-efficient mixing.

Furthermore, other studies have also been carried out focusing on the analysis of mixing efficiency using non-Newtonian fluids. Montante et al. [33] conducted experimental and CFD study to evaluate the influence of tracer injection positions, turbulence models, and numerical parameters on the predicted mixing time in baffled stirred tanks using Newtonian and pseudoplastic liquids. The tank was stirred using multiple down-pumping, 45-degree, pitched-blade turbine (PBT). The mixing time calculation was based on the tracer concentration within the stirred vessel. The results of the investigation showed that the tracer concentration dynamics in both the experimental measurements and CFD results obtained for a variety of conditions does not show any significant difference. Kazemzadeh et al. [34] performed a similar study on Herschel–Bulkley fluids stirred with coaxial mixers. Mixing time measurements were performed on opaque xanthan gum solution by means of electrical resistance tomography (ERT). The result of the study confirms that mixing time and power consumption increased with the fluid consistency index and yield stress and decreased with an increase in the power-law index. Subsequently, design of experiments (DoE) and response surface methodology (RSM) analysis demonstrated that the consistency index and speed ratio are the most significant factors that influence the mixing efficiency.

The main aim of this chapter is to demonstrate the practical implementation of the numerical analysis of mixing performance in a stirred tank reactor (STR) and how to parametrise and sequentially optimise the various parameters that drive the process using a commercial CFD software code. Most of the simulation workflow implemented in this chapter can also be performed using any open-source numerical code. However, it is worth pointing out that a good knowledge of the process being modelled and the underlying mechanistic assumptions that govern the process, as well as a good mastery of the CFD code, are highly indispensable to achieve accurate and reproducible results.

5.5 NUMERICAL SIMULATION OF HYDRODYNAMICS OF STIRRED TANK

5.5.1 Governing Equations of Fluid Flow and Specie Transport

The single-phase fluid flow in a stirred tank for an incompressible fluid is governed by the conservation equations of mass and momentum along with the specie transport equations for the tracer concentration under turbulent conditions. The governing equations expressed mathematically are given in Equations (5.9)–(5.12). A full derivation of these governing equations and detailed guidance on their specific applications can be found in the ANSYS Fluent theory guide and other publications on fluid dynamics [35, 36]:

$$\frac{\partial v_i}{\partial x_i} = 0 \tag{5.9}$$

$$\rho \frac{\partial \overline{v}_i}{\partial t} + \rho \overline{v} \frac{\partial \overline{v}_i}{\partial x_j} = -\frac{\partial \overline{p}}{\partial x_i} + \frac{\partial}{\partial x_j}\left(\overline{\tau}_{ij} - \rho \overline{v_i' v_j'}\right) + \rho \vec{g} \tag{5.10}$$

$$\rho \frac{\partial Y_i}{\partial t} + \rho \overline{v} \frac{\partial Y_i}{\partial x_j} = -\frac{\partial}{\partial x_j}\left(\vec{J}_i\right) + S_i \tag{5.11}$$

$$\vec{J}_i = \left(\rho D_{i,m} + \frac{\mu_t}{Sc_t}\right)\frac{\partial Y_i}{\partial x_j} \tag{5.12}$$

In Equations (5.9)–(5.12), ρ is the fluid density, $\bar{v}$ is the mean velocity, $\bar{p}$ is the mean pressure, $\bar{\tau}_{ij}$ is the viscous stress tensor, $\rho \overline{v_i' v_j'}$ is the Reynolds stresses which represent the transfer of momentum due to turbulent fluctuations, t is the time, $\vec{g}$ is the gravitational body force acting in the negative y direction, S_i is the source term due to the rate of creation by addition from dispersed phase to the continuous phase, Y_i is the mass fraction of each specie in the mixture, $\vec{J_i}$ is the turbulent mass diffusion flux of each specie, $D_{i,m}$ is the turbulent mass diffusion coefficient, μ_t is the turbulent viscosity, and Sc_t is the turbulent Schmidt number.

5.5.2 Description of the Geometry and Flow Domain

The governing equations are solved numerically on a computational grid using a commercial CFD solver (ANSYS Fluent v2025R1). Numerical simulation was performed by creating a 3D geometry and mesh of a 733-litre cone-shaped mixing tank with a diameter of 0.597 m and a height of 0.8 m using built-in ANSYS solid modelling and meshing software. The computational grid consists of high-quality hexahedral and polyhedral meshes. The multiple reference frame (MRF) approach was employed to mimic the rotation of a two-blade axial flow impeller, which acts as the stirrer and flow generator. The stirrer was placed with a clearance of 0.3 m from the base of the tank. The mixing tank stirrer-vessel configuration as well as the tracer injection points and positions of the probes for tracer detection is shown in Figure 5.5.

5.5.3 Numerical Solution of the Governing Equations

The mixing process was simulated in batch mode with the mixing tank completely filled with water as the working fluid at room temperature (20°C) and pressure (1,01,325 Pa) and operated over a range of different rotational speeds, blade chord, baffle length, and thickness. Additionally, a fluid–structure interaction (FSI) analysis was also performed to evaluate the hydrodynamic stress on the impeller blade during the mixing process. Reynolds-averaged Navier–Stokes (RANS) simulations were performed using the Renormalisation Group RNG $k - \varepsilon$ turbulence model to account for the swirling effect of the turbulent motion of the flow, especially near the impeller. A density-based solver is used for the velocity–pressure

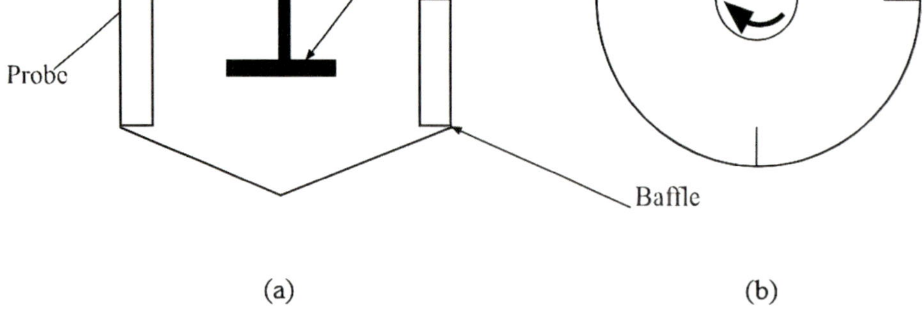

FIGURE 5.5 Schematic of the reactor geometry and flow domain (not to scale): (a) side view and (b) top view.

coupling while the least square cell-based method was employed for the gradient. Additionally, appropriate numerical settings such as SIMPLEC for the pressure–velocity coupling and the second-order upwind spatial discretisation scheme for all the flow variables, with the exception of pressure for PRESTO, are used. The second-order implicit temporal discretisation scheme was used for the time advancement in the transient analysis of the tracer transport after the steady-state fluid flow solution. In addition, material properties of the fluid enclosed within the reactor were specified using the built-in Fluent materials database while the tracer properties were set to match those of the water. The solution residuals, moment, average kinetic energy, and tracer mass fraction are used as the convergence parameters for the numerical solution, while the solution is considered to have converged when the residual reduces to an acceptable level of 1×10^{-3}. All numerical simulations were performed on a laptop computer with a quad-core i5-6500, 3.20 GHz 6MB cache CPU with 32 GB RAM.

5.5.4 Analysis of Mixing Performance

Mixing analysis was carried out by simulating tracer dispersion within the mixing tank. This was achieved by performing a steady-state numerical simulation of fluid flow after which the flow equations are switched off and the tracer is injected from the fluid surface at the top of the tank, and a transient analysis is performed to monitor the dispersion of the mixer in the tank. The tracer is treated as a passive scalar as it does not interfere with the fluid flow and its material properties were specified to match those of the fluid. The dispersion of the tracer was monitored at several probe locations in the tank by calculating the change in the tracer concentration at these points as the simulation progressed after which mixing profiles were plotted to estimate the local mixing time at these points. The global mixing time for the stirred tank was determined by calculating the mixing degree (spatial variance of the tracer concentration) given by Equation (5.13) with respect to time [37, 38]. The time taken to reach an equilibrium or a completely uniform mixture (100% homogeneity) for all the probing locations is the time at which the computed degree of mixing is equal to zero:

$$C_{mix} = \sqrt{\frac{\Sigma_i \left(\frac{c_i - \overline{c}}{\overline{c}} \right) \Delta V_i}{\Sigma_i \Delta V_i}} \tag{5.13}$$

In Equation (5.13), C_{mix} is the degree of mixing, i is the location of monitoring point, V_i is the cell volume at the monitoring point i, c_i is the tracer concentration at monitoring point i, $\overline{c}$ is the equilibrium tracer concentration at complete homogenisation.

5.6 RESULTS AND DISCUSSION

5.6.1 Steady-State Mean Flow Characteristics

The mean flow characteristics at the steady state condition across the vertical and horizontal planes are shown in Figure 5.6. It can be observed from Figure 5.6a–5.6d that the axial pumping effect is highest close to the impeller with a circulating vortex on either side of the impeller. The general flow field is qualitatively well described with the highest velocity magnitude as expected occurring close to the impeller blade tips. Figure 5.6e, f provides a quantitative description of the average flow parameters. Figure 5.6e shows

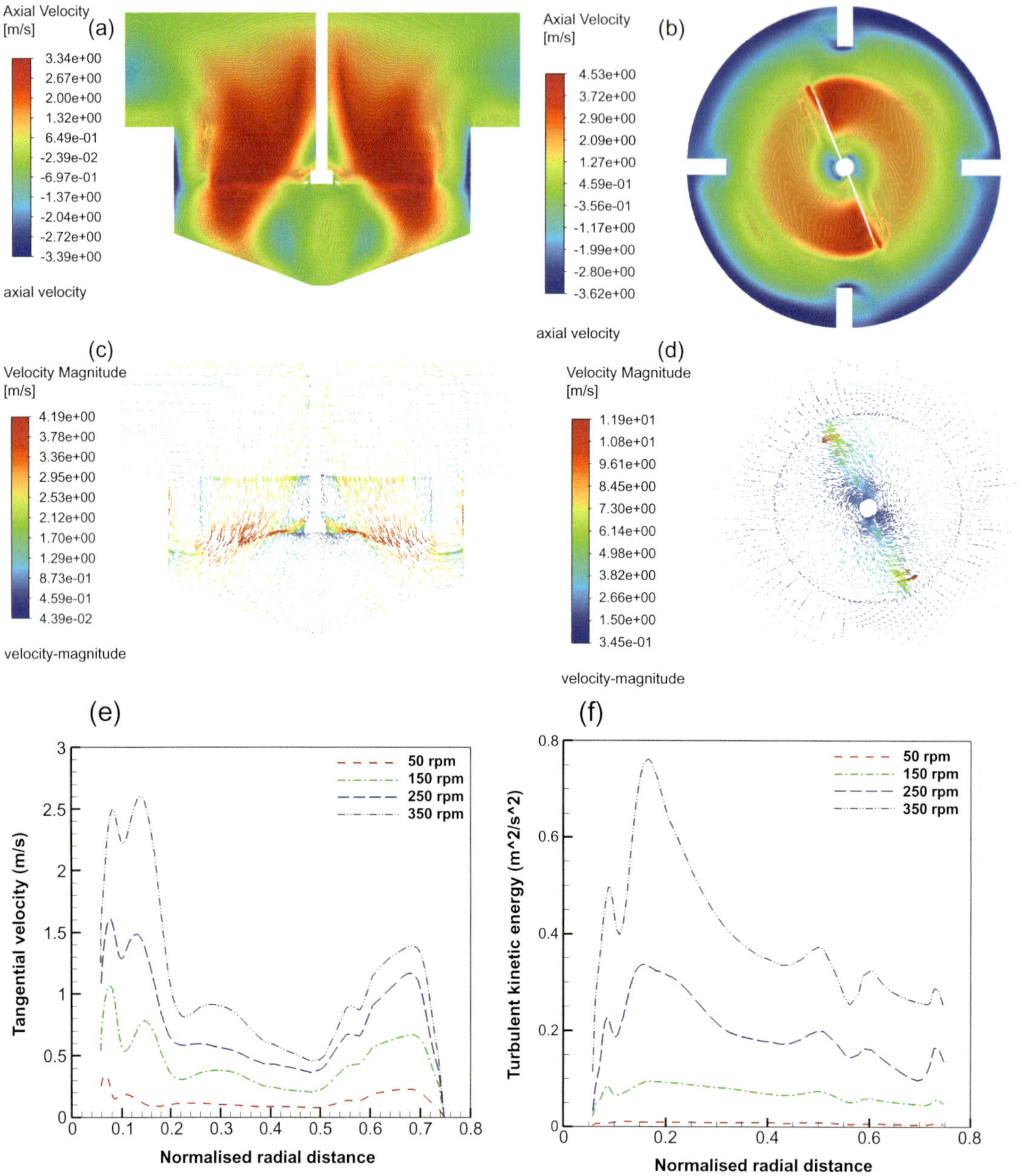

FIGURE 5.6 Steady-state mean flow parameters along vertical *XY* and horizontal *XZ* planes: (a) vertical plane axial velocity contour at 250 RPM, (b) horizontal plane axial velocity contour at 250 RPM, (c) vertical plane velocity vector at 250 RPM, (d) horizontal plane velocity contour at 250 RPM, (e) horizontal plane tangential velocity profiles, and (f) horizontal plane turbulent kinetic energy profiles.

the tangential velocity variation along the centre line of a horizontal plane near the impeller. It can be observed that the tangential velocities increase with the rotation speed, with the highest values recorded close to the shaft of the impeller. Figure 5.6f shows a similar quantitative description for the turbulent kinetic energy on the same horizontal plane. The trend is seen to be consistent with the tangential velocity as the kinetic energy also increases with impeller rotational speed and is highest close to the impeller and reduces substantially towards the tank wall.

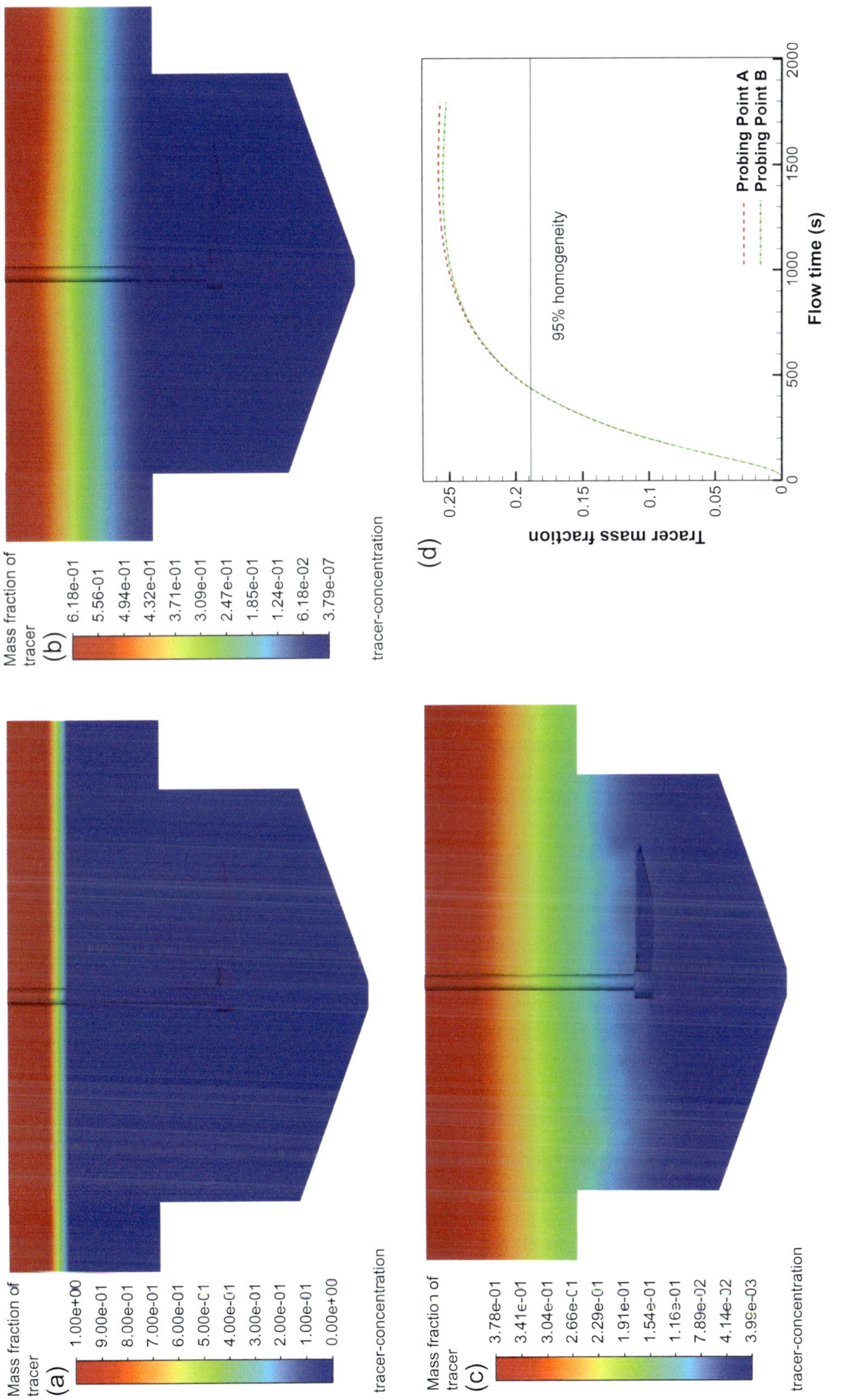

FIGURE 5.7 Evolution of the tracer concentration in the mixing tank: (a) $t = 1$ s, (b) $t = 600$ s, (c) $t = 1800$ s, and (d) tracer concentration response curves at two probing points after 1800 s at a rotational speed of 350 RPM.

5.6.2 Spatiotemporal Evolution of the Tracer Concentration

Mixing efficiency is an important performance criterion for most practical applications of stirred tank reactors (STRs). A good mixing strategy ensures the best possible mixing efficiency at the lowest energy consumption. In order to get an insight into the mixing process, transient simulations were performed by injecting tracer into the mixing tank at the top layer and thereafter monitor the diffusion of the tracer as the fluid is stirred. Mixing performance was assessed in terms of the tracer concentration at several probing locations within the tank. The spatiotemporal evolution of the tracer dispersity is shown in Figure 5.7a–c while the tracer concentration profiles are presented in Figure 5.7d, which clearly show a sustained change in the tracer dispersity with time. It can be seen from Figure 5.7a–c that there is a sustained diffusion of the tracer as the simulation progresses with the process slowing down as the tracer concentration approaches equilibrium values for the mixing tank, and this is consistent with other studies reported elsewhere [35]. The concentration profiles also show that the tracer concentration at the two observation points levels off as the concentration approaches the volume-averaged values for the whole mixing tank. The local mixing time at the observation points was adjudged to be approximately 480 s, which is the point at which concentration is equal to 95% of the volume-averaged value for the whole tank.

5.6.3 Optimisation of the Reactor Design and Process Parameters

One of the primary goals of any mixing process is to achieve an acceptable level of mixing performance with minimal energy consumption. In order to meet this goal, a parametric study was performed by following the approach described in studies conducted by Mukhtar et al. and Golzarijalal et al. [39, 40]. The study provides an evaluation of the influence of the mixing tank design factors (baffle length, baffle thickness, and blade chord) and operating parameter (impeller speed) and their interactions on the overall process performance such as the power consumption, average kinetic energy, and impeller torque (responses). A design of experiment (DoE) table that covers the design space was generated using central composite design (CCD) showing all possible combinations of the factors and responses. Numerical experiments consisting of 26 runs were randomly performed at these design points to generate the response results from which a Pareto chart was created that shows the effects of the design factors and their interactions on the responses. Thereafter, a regression analysis was performed by fitting a second-degree polynomial function and a Gaussian process regression to the statistically significant factors to predict the responses. Finally, design optimisation was carried out to determine optimum combinations of various parameters that influence the mixing performance.

5.6.3.1 Screening of Design Factors Using Pareto Analysis

Pareto analysis was carried out to determine the design factors and their interactions that strongly influence the reactor's operating parameters. Pareto charts shown in Figure 5.8 were employed to identify the main and the interactive parameters with statistically significant effect on the responses. This was done by exporting the design of experiments table from ANSYS DesignXplorer to Statistical software (TIBCO software Inc.) where the Pareto charts were generated by considering the effects of the individual factors and their interactions on the responses. Factors and their interactions in the Pareto chart that exceed the reference line in red colour are considered to have significant effects on the responses at a confidence level of 95%. Figure 5.8a shows that the impeller speed is the main parameter that has the strongest effect on the average kinetic energy followed by the baffle length and blade chord. In addition, the average kinetic energy is also sensitive to the interactions between the impeller speed and baffle length as well as its interactions with the blade chord. Figure 5.8b shows that power consumption is only affected by impeller speed and baffle length and its interactions. In the case of impeller torque shown in Figure 5.8c, the significant

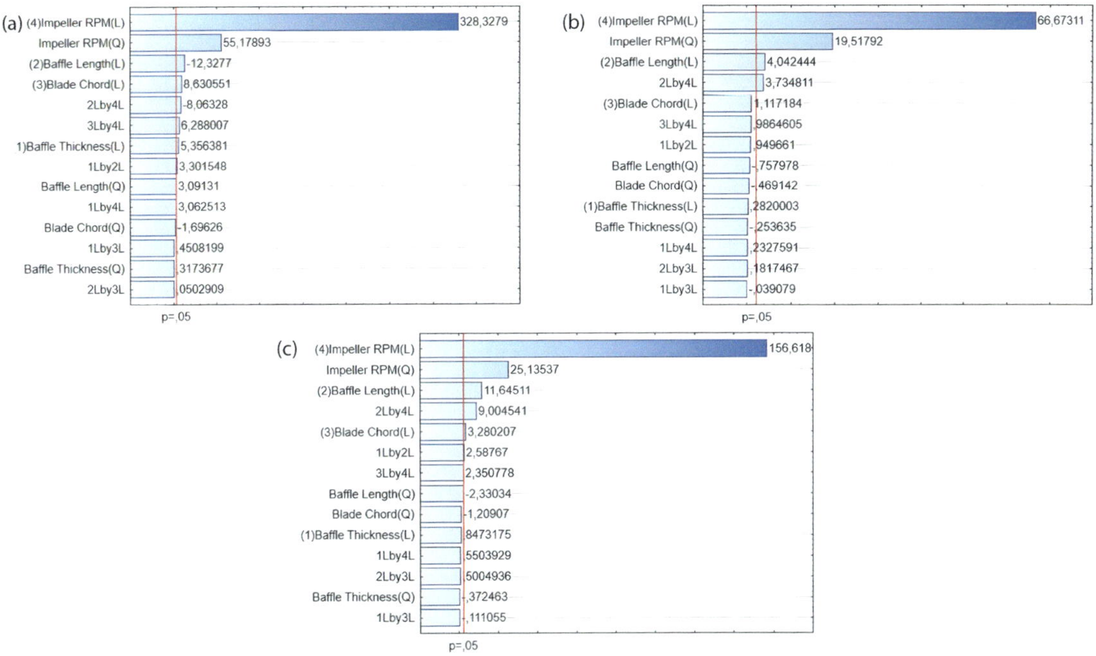

FIGURE 5.8 Pareto chart showing the effect of the design factors and their interactions on the operating parameters (responses): (a) average kinetic energy, (b) power consumption, and (c) impeller torque.

factors are the impeller speed, baffle length, and blade chord, as well as the interaction between impeller speed and baffle length.

5.6.3.2 Analysis of Response Surface and Contour Charts

The Pareto charts shown in the earlier section revealed that the impeller speed, baffle length, blade chord, and the interactions between them affect the response parameters such as the power consumption, average kinetic energy, and impeller torque. In addition to the Pareto charts, surface and contour plots were generated using Statistica software to visualise and illustrate these effects on the various responses with a view to optimising these factors to obtain the desired responses. Figure 5.9a, b shows the interactive effect of baffle length and impeller speed as well as that of blade chord and impeller speed on the average kinetic energy. The figures reveal that the average kinetic energy increases with impeller speed but slightly decreases with increasing baffle length. Also, it slightly increases as the blade chord increases, especially at an impeller speed of above 60 RPM. The reduction in impeller speed is due to increased turbulence as the baffle length increases which reduces the impeller rotation, while an increase in the blade chord results in an increase in stirring intensity of the blade, which results in an increase in the average kinetic energy. Figure 5.9c illustrates the interactive effect of the baffle length and impeller speed on the power consumption. It shows that the power consumption increases as expected with the impeller rotational speed, especially at an impeller speed of above 60 RPM due to the additional power required to make the impeller rotate faster. In the case of the baffle length, the power consumption increases slightly with the increase in baffle length and then levels off. This is due to the fact that the increase in baffle length promotes mixing up to a point after which any further increase in the baffle length will not yield additional benefit in terms of mixing efficiency. As longer baffles create resistance to fluid motion, this makes the impeller draw more power to maintain the desired mixing efficiency. Figure 5.9d shows the effect of baffle length and impeller speed (rpm) on the impeller torque. It can be observed that the impeller torque increases as expected with an increase in the impeller speed due to the higher rotational force. Also, the impeller speed increases with an increase in baffle length, especially at RPM values higher than

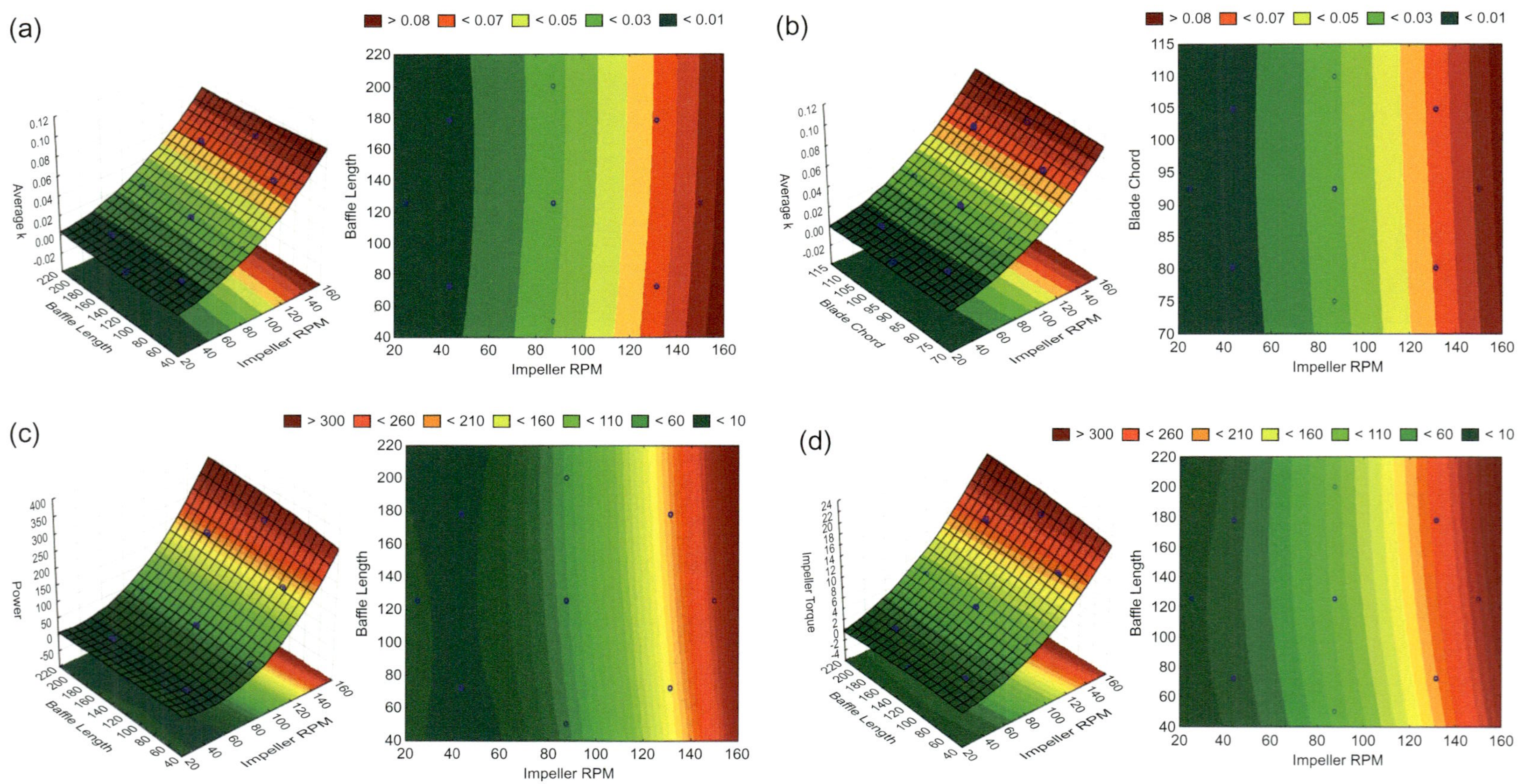

FIGURE 5.9 Response surface and contour plots for the polynomial regression showing the response parameters as a function of two input factors while other factors are held constant at their respective centre points: (a) average kinetic energy as a function of baffle length and impeller speed, (b) average kinetic energy as a function of blade chord and impeller speed, (c) power consumption as a function of baffle length and impeller speed, and (d) impeller torque as a function of baffle length and impeller speed.

TABLE 5.1 Range of Design Factors and Responses and the Optimisation Strategy

	FACTORS					*RESPONSES*	
PARAMETERS	*BAFFLE THICKNESS (MM)*	*BAFFLE LENGTH (MM)*	*BLADE CHORD (MM)*	*IMPELLER SPEED (RPM)*	*AVERAGE KE ($JKGS^{-1}$)*	*IMPELLER TORQUE (J)*	*POWER ($JKGM^2S^{-3}$)*
Lower bound	10	50	75	25	—	—	—
Upper bound	50	200	110	150	—	—	—
Calculated minimum	—	—	—	—	0.0014	0.33	0.8635
Calculated maximum	—	—	—	—	0.088	19.79	310.79
Target and constraints	27.11	50.32	77.73	91.51	≥0.03	—	Minimise

TABLE 5.2 Optimised Design Points and Calculated Responses for the Polynomial Regression

PARAMETERS	*FACTORS*					*RESPONSES*	
CANDIDATE POINTS	*BAFFLE THICKNESS (MM)*	*BAFFLE LENGTH (MM)*	*BLADE CHORD (MM)*	*IMPELLER SPEED (RPM)*	*AVERAGE KE ($JKGS^{-1}$)*	*IMPELLER TORQUE (J)*	*POWER ($JKGM^2S^{-3}$)*
1	16.80	61.22	77.76	90.36	0.030	5.73	54.99
2	20.86	54.91	78.02	90.89	0.031	5.85	55.19
3	27.11	50.32	77.73	91.51	0.032	5.87	55.35

100. This increase in torque at high baffle length can be attributed to the interactive effect of baffle length and impeller RPM. Given that an increase in baffle length increases turbulence which slows down fluid motion, it seems that this is not a limiting factor as the impeller RPM values increase above a particular threshold.

5.6.3.3 *Optimisation Strategy and Optimal Design Points*

The goal of the design optimisation was to minimise the power consumption of the impeller, while being constrained by the minimum values for the average kinetic energy. This goal was achieved by defining an objective function in ANSYS DesignXplorer for the response parameters and thereafter optimising the defined functions and constraints using a multi-objective genetic optimisation algorithm (MOGA), which identifies the best combinations of the input factors that satisfy the desired responses. MOGA is a variant of the popular NSGA-II (non-denominated sorted genetic algorithm), which supports multiple objectives and constraints and aims at finding the global optimum. Table 5.1 shows the range of design points for the design factors and the calculated values of the responses, while Table 5.2 shows the optimised candidate design points. It can be seen from the table that three design options have been generated, which satisfies the defined objective functions. It is worth pointing out that the data analysis and optimisation workflow can also be implemented using MATLAB® or Python software.

5.6.4 Towards a Data-Driven Future

The use of machine learning and other data modelling and optimisation tools in the modelling and optimisation of various environmental processes has grown exponentially over the last few decades. These data-driven models which can be applied to both experimental and computational data provide a fast and flexible approach to getting insight from a large pool of data such as those generated from computational experiments [41, 42]. Surrogate models such as the polynomial regression, Gaussian Process Regression or Kriging and artificial neural networks (ANNs) are by far the most popular machine learning models employed in the optimisation of physicochemical processes in the environmental engineering field due to their ability to reproduce and model nonlinear relationships [43, 44]. The use of advanced surrogate models and neural networks as metamodeling tools for fitting response surfaces has been widely employed in many scientific studies. Neural networks has been shown to show superior performance when compared to other classical response surface models especially in a situation where the response is not 'smooth' [39].

A new frontier in the field of mathematically modelling is the hybrid data-driven modelling of engineering problems, which is fast evolving as a distinct field of research with huge potential. This approach can be employed either as a low-fidelity alternative to time-consuming and expensive numerical simulation of mixing tanks and process reactors or as a viable replacement for the existing modelling tools [45]. While the scientific machine learning field of research is still at its infancy, a number of promising research studies have been published demonstrating how this approach can be applied to solve complex engineering fluid flow problems [46–50]. Most of the reported studies in this field are either based on the classical surrogate models (polynomial, Kriging, KNN etc.), reduced order model (ROM) or, more recently, the so-called physics-informed neural networks (PINNs) in which information about the governing equations of the physical problem and the boundary conditions is embedded in the loss functions of the neural networks. These enhanced neural networks are able to learn the underlying complex physics involved in a physical problem and make better predictions. The greatest challenge, especially in the modelling of complex process engineering fluid flow problems, is the high computational cost associated with the initial training of these neural networks especially when temporal information is included as an input in the neural networks. As progress is made in this field and more robust and highly efficient PINNs are developed, it is anticipated that the scientific machine learning algorithms will play a more crucial role in engineering simulations. With the rapid growth in computational capabilities, these techniques are expected to become standard tools in intelligent design and optimisation of stirred tank reactors in the future.

5.7 CONCLUDING REMARKS

This chapter presents and discusses theoretical and numerical approaches to the analysis of mixing processes in stirred tanks. The numerical simulation of the fluid flow and mixing was performed on a 733 L tank using a commercial software code. It is worth pointing out that the result from this numerical study needs to be validated by performing additional experimental measurements such as the particle image velocimetry (PIV) and colorimetry to quantify the flow parameters and the tracer concentration. The aim of the CFD analysis carried out in this chapter was to demonstrate the application of theoretical knowledge and numerical software code as a study and research tool in the analysis of mixing tank and multiphase flow problems in environmental engineering. The following conclusions can be drawn from this case study:

- The concept of fluid mixing and various length scales involved in the mixing process has been discussed from a theoretical perspective.
- The stirrer-vessel concept and hydrodynamic characterisation of fluid flow have been discussed from both empirical and theoretical perspectives.
- The governing equations of fluid flow and specie transport for fluid flows typically encountered in stirred tank reactors have been presented.
- The workflow and numerical simulation strategy for a successful implementation of numerical solution for typical mixing problems have been presented and discussed.

- Finally, numerical simulation on a case study of mixing tank has been carried out to characterise fluid flow parameters and mixing performance.

DATA AVAILABILITY

The data that support the findings of this chapter are available from the corresponding author, [B.O.], upon reasonable request.

NOMENCLATURE

Latin Letters

Variable	Description, Units
d	**Diffusion coefficient, —**
v'	**Velocity fluctuation, m/s**
N_p	**Power number, —**
N_Q	**Flow number, —**
M_O	**Momentum number, —**
P	**Impeller power consumption, W**
N	**Impeller rotational speed, rpm**
D	**Impeller diameter, mm**
$M_{\theta 95}$	**Mixing time at 95% homogeneity, s**
T	**Mixing tank diameter, m**
H	**Tank liquid height, m**
Q	**Impeller pumping capacity, m^3/s**
C_{mix}	**Degree of mixing, —**
V_i	**Cell volume at monitoring point, m^3**
c_i	**Tracer concentration at monitoring point *i*, —**
$\overline{c}$	**Equilibrium tracer concentration at complete homogenisation, —**
Y_i	**Mass fraction of each specie, —**
$\vec{J}_i$	**Turbulent mass diffusion flux, $kg/m^2\ s$**
$D_{i,m}$	**Turbulent mass diffusion coefficient, m^2/s**
Sc_t	**Turbulent Schmidt number, —**

Greek Symbols

Variable	Description, Units
λ_0	**Kolmogorov length scale, —**
λ_B	**Batchelor length scale, —**
λ_T	**Taylor length scale, —**
Λ	**Integral length scale, —**
ε	**Mean dissipation rate of kinetic energy, m^2/m^3**

ρ	**Fluid density, kg/m³**
v	**Kinematic viscosity, m²/s**
$\overline{v}$	**Mean velocity, m/s**
$\overline{p}$	**Mean pressure, Pa**
$\overline{\tau}_{ij}$	**Viscous stress tensor, N/m²**
$\rho\overline{v_i'v_j'}$	**Reynolds stress, N/m²**
$\vec{g}$	**Gravitational body force, m/s²**
μ_t	**Turbulent viscosity, Pa s**

Acronyms

CFD	Computational fluid dynamics
DES	Detached eddy simulation
DoE	Design of experiments
MOGA	Multi-objective genetic algorithm
NASA	Non-denominated sorted genetic algorithm
RT	Rushton turbine
PBT	Pitched-blade turbine
STR	Stirred tank reactor
RANS	Reynolds-averaged Navier–Stokes
MRF	Multiple reference frame
CCD	Central composite design
ROM	Reduced order model
PINN	Physics-informed neural network

APPENDIX

A.5.1 Design of Experiments (DoE) and Optimised Design Point Tables

TABLE A.5.1 Central Composite Design (DoE) Table for Response Surface Analysis

PARAMETERS	*FACTORS*				*RESPONSES*		
CANDIDATE POINTS	*BAFFLE THICKNESS (MM)*	*BAFFLE LENGTH (MM)*	*BLADE CHORD (MM)*	*IMPELLER SPEED (RPM)*	*AVERAGE KE (JKGS⁻¹)*	*IMPELLER TORQUE (J)*	*POWER (JKGM²S⁻³)*
1	30	125	92.50	87.50	0.028	6.22	57.01
2	10	125	92.50	87.50	0.026	6.13	56.11
3	50	125	92.50	87.50	0.029	6.24	57.22
4	30	50	92.50	87.50	0.030	5.43	49.73
5	30	200	92.50	87.50	0.027	6.46	59.18
6	30	125	75.00	87.50	0.026	5.90	54.08
7	30	125	110.00	87.50	0.028	6.26	57.36
8	30	125	92.50	25.00	0.002	0.48	1.27

(*Continued*)

TABLE A.5.1 Continued

PARAMETERS	FACTORS				RESPONSES		
CANDIDATE POINTS	BAFFLE THICKNESS (MM)	BAFFLE LENGTH (MM)	BLADE CHORD (MM)	IMPELLER SPEED (RPM)	AVERAGE KE (JKGS^{-1})	IMPELLER TORQUE (J)	POWER (JKGM^{2}S^{-3})
9	30	125	92.50	150.00	0.082	18.17	285.45
10	15.92	72.19	80.18	43.49	0.007	1.39	6.31
11	44.08	72.19	80.18	43.49	0.007	1.37	6.23
12	15.92	72.19	80.18	43.49	0.006	1.51	6.83
13	44.08	72.19	80.18	43.49	0.006	1.56	7.11
14	15.92	72.19	104.82	43.49	0.007	1.43	6.53
15	44.08	72.19	104.82	43.49	0.007	1.41	6.41
16	15.92	177.82	104.82	43.49	0.006	1.57	7.16
17	44.08	177.82	104.82	43.49	0.007	1.62	7.37
18	15.92	177.82	80.18	131.51	0.064	12.64	174.11
19	44.08	177.82	80.18	131.51	0.064	12.38	170.51
20	15.92	177.82	80.18	131.51	0.059	13.91	191.53
21	44.08	177.82	80.18	131.51	0.061	14.42	198.60
22	15.92	72.19	104.82	131.51	0.066	13.07	179.96
23	44.08	72.19	104.82	131.51	0.067	12.72	175.15
24	15.92	177.82	104.82	131.51	0.062	14.44	198.81
25	44.08	177.82	104.82	131.51	0.064	14.97	206.16

TABLE A.5.2 Optimised Design Points and Calculated Responses for the Gaussian Process Regression (Kriging)

PARAMETERS	FACTORS				RESPONSES		
CANDIDATE POINTS	BAFFLE THICKNESS (MM)	BAFFLE LENGTH (MM)	BLADE CHORD (MM)	IMPELLER SPEED (RPM)	AVERAGE KE (JKGS^{-1})	IMPELLER TORQUE (J)	POWER (JKGM^{2}S^{-3})
1	35.92	50.32	79.14	88.47	0.029	5.18	47.97
2	35.80	50.30	89.00	88.19	0.030	5.30	48.97
3	35.87	51.64	93.19	87.97	0.030	5.34	49.16

REFERENCES

1. V. Buwa, A. Dewan, A.F. Nassar, F. Durst, Fluid dynamics and mixing of single-phase flow in a stirred vessel with a grid disc impeller: Experimental and numerical investigations, *Chem. Eng. Sci.* 61 (2006) 2815–2822. https://doi.org/10.1016/j.ces.2005.10.066.
2. B. Oyegbile, G. Akdogan, M. Karimi, Experimental and CFD studies of the hydrodynamics in wet agglomeration process, *ChemEngineering* 2 (2018) 32–49. https://doi.org/10.3390/chemengineering2030032.
3. L. Falk, J.-M. Commenge, Characterization of mixing and segregation in homogeneous flow systems, in: V. Hessel, A. Renken, J.C. Schouten, J.-I. Yoshida (eds.), *Handbook of Micro Reactors*, John Wiley & Sons, Weinheim, 2009: pp. 147–171. https://doi.org/10.1002/9783527631445.ch6.

4. D.H. Bache, R. Gregory, *Flocs in Water Treatment*, IWA Publishing, London, 2007. https://doi.org/10.2166/9781780402000.
5. J. Bridgeman, B. Jefferson, S.A. Parsons, Computational fluid dynamics modelling of flocculation in water treatment: A review, *Eng. Appl. Comput. Fluid Mech.* 3 (2009) 220–241. https://doi.org/10.1080/19942060.2009.11015267.
6. K. Mühle, Floc stability in laminar and turbulent flow, in: B. Dobias (ed.), *Coagulation and Flocculation: Theory and Applications*, Marcel Dekker, New York, NY, 1993: pp. 355–390.
7. M.M. Benjamin, D.F. Lawler, *Water Quality Engineering: Physical/Chemical Treatment Processes*, John Wiley & Sons, Hoboken, NJ, 2013.
8. D.H. Bache, R. Gregory, *Flocs in Water Treatment*, IWA Publishing, London, 2007.
9. J. Bałdyga, J.R. Bourne, A fluid mechanical approach to turbulent mixing and chemical reaction part II micromixing in the light of turbulence theory, *Chem. Eng. Commun.* 28 (1984) 243–258. https://doi.org/10.1080/00986448408940136.
10. S.F. Thomas, P. Rooks, F. Rudin, N. Cagney, S. Balabani, S. Atkinson, P. Goddard, R.M. Bransgrove, P.T. Mason, M.J. Allen, Swirl flow bioreactor containing dendritic copper-containing alginate beads: A potential rapid method for the eradication of *Escherichia coli* from waste water streams, *J. Water Process Eng.* 5 (2015) 6–14. https://doi.org/10.1016/j.jwpe.2014.10.010.
11. S.M. Kresta, R.S. Brodkey, Turbulence in mixing applications, in: E.L. Paul, V.A. Atiemo-Obeng, S.M. Kresta (eds.), *Handbook of Industrial Mixing: Science and Practice*, John Wiley & Sons, Hoboken, NJ, 2004: pp. 19–87. https://doi.org/10.1002/0471451452.ch2.
12. M. Zlokarnik, *Stirring: Theory and Practice*, Wiley-VCH, Weinheim, 2008.
13. D. Thoenes, *Chemical Reactor Development: From Laboratory Synthesis to Industrial Production*, Springer, Dordrecht, 1998.
14. T. Sparks, *Fluid Mixing in Rotor/Stator Mixers*, PhD Thesis, Cranfield University, 1996.
15. J. Bałdyga, R. Pohorecki, Turbulent micromixing in chemical reactors—A review, *Chem. Eng. J. Biochem. Eng. J.* 58 (1995) 183–195. https://doi.org/10.1016/0923-0467(95)02982-6.
16. J.Y. Oldshue, R.R. Trussell, Design of impellers for mixing, in: A. Amirtharajah, M.M. Clark, R. Trussell (eds.), *Mixing in Coagulation and Flocculation*, American Water Works Association, Denver, CO, 1991: pp. 309–342.
17. B.E. Logan, *Environmental Transport Processes*, John Wiley & Sons, Hoboken, NJ, 2012.
18. K.A. Kruster, *The Influence of Turbulence on Aggregation of Small Particles in Agitated Vessels*, PhD Thesis, Technische Universiteit Eindhoven, 1991.
19. N. Kockmann, *Transport Phenomena in Micro Process Engineering*, Springer, Heidelberg, 2008.
20. F. Maggi, *Flocculation Dynamics of Cohesive Sediment*, PhD Thesis, Delft University of Technology, 2005.
21. J. Bałdyga, J.R. Bourne, *Turbulent Mixing and Chemical Reactions*, Wiley-VCH, Weinheim, 1999.
22. N. Harnby, M.F. Edwards, A.W. Nienow (eds.), *Mixing in the Process Industries*, Butterworth-Heinemann, Oxford, 1997. https://doi.org/10.1016/B978-0-7506-3760-2.X5020-3.
23. E.L. Paul, V.A. Atiemo-Obeng, S.M. Kresta (eds.), *Handbook of Industrial Mixing: Science and Practice*, John Wiley & Sons, Hoboken, NJ, 2004. https://doi.org/10.1002/0471451452.
24. J.C. Godfrey, A. Amirtharajah, Mixing in liquids, in: A. Amirtharajah, M.M. Clark, R. Trussell (eds.), *Mixing in Coagulation and Flocculation*, American Water Works Association, Denver, CO, 1991: pp. 35–79.
25. M.B. Machado, J.R. Nunhez, D.N., S.M. Kresta, Impeller characterization and selection: Balancing efficient hydrodynamics with process mixing requirements, *AIChE J.* 58 (2012) 2573–2588. https://doi.org/10.1002/aic.12758.
26. R.R. Hemrajani, G.B. Tatterson, Mechanically stirred vessels, in: E.L. Paul, V.A. Atiemo-Obeng, S.M. Kresta (eds.), *Handbook of Industrial Mixing: Science and Practice*, John Wiley & Sons, Hoboken, NJ, 2004: pp. 345–390. https://doi.org/10.1002/0471451452.ch6.
27. Ö. Aydin, M.E. Arzutuğ, S. Yapici, A new application area of electrochemical limiting diffusion current technique: Measurement of mixing time and effect of some parameters, *SN Appl. Sci.* 1 (2019). https://doi.org/10.1007/s42452-019-1378-3.
28. G. Bågmark, *Mixing Time Prediction in Stirred Tanks Using Empirical Methods and Computational Fluid Dynamics*, MSc Thesis, Lund University, 2024. https://lup.lub.lu.se/student-papers/search/publication/9162472.
29. H. Kramers, G.M. Baars, W.H. Knoll, A comparative study on the rate of mixing in stirred tanks, *Chem. Eng. Sci.* 2 (1953) 35–42. https://doi.org/10.1016/0009-2509(53)80006-0.
30. R. Afshar Ghotli, M.S. Shafeeyan, M.R. Abbasi, A.A. Abdul Raman, S. Ibrahim, Macromixing study for various designs of impellers in a stirred vessel, *Chem. Eng. Process. Process Intensif.* 148 (2020) 107794. https://doi.org/10.1016/j.cep.2019.107794.

31. Z. Yao, H. Xu, J. Li, T. Xu, Multi-objective optimization of stirring tank based on multiphase flow simulation, *Chem. Eng. Res. Des.* 189 (2023) 680–693. https://doi.org/10.1016/j.cherd.2022.11.043.
32. F.L. Yang, Turbulent flow and mixing performance of a novel six-blade grid disc impeller, *Korean J. Chem. Eng.* 32 (2015) 816–825. https://doi.org/10.1007/s11814-014-0255-4.
33. G. Montante, M. Moštěk, M. Jahoda, F. Magelli, CFD simulations and experimental validation of homogenisation curves and mixing time in stirred Newtonian and pseudoplastic liquids, *Chem. Eng. Sci.* 60 (2005) 2427–2437. https://doi.org/10.1016/j.ces.2004.11.020.
34. A. Kazemzadeh, F. Ein-Mozaffari, A. Lohi, L. Pakzad, Effect of the rheological properties on the mixing of Herschel-Bulkley fluids with coaxial mixers: Applications of tomography, CFD, and response surface methodology, *Can. J. Chem. Eng.* 94 (2016) 2394–2406. https://doi.org/10.1002/cjce.22601.
35. Y. Zhou, L. Wang, H. He, J. Sun, Mixing process in a tank stirred with improved double intermig impellers, *J. Chem. Eng. Japan* 52 (2019) 719–729. https://doi.org/10.1252/jcej.18we236.
36. A. Issakhov, A.R. Baitureyeva, Numerical modelling of a passive scalar transport from thermal power plants to air environment, *Adv. Mech. Eng.* 10 (2018). https://doi.org/10.1177/1687814018799544.
37. L. Li, B. Xu, CFD simulation of local and global mixing time in an agitated tank, *Chin. J. Mech. Eng.* 30 (2017) 118–126. https://doi.org/10.3901/CJME.2016.1107.129.
38. M.F. Edwards, M.R. Baker, J.C. Godfrey, Mixing of liquids in stirred tanks, in: N. Harnby, M.F. Edwards, A.W. Nienow (eds.), *Mixing in the Process Industries*, Butterworth-Heinemann, Oxford, 1997: pp. 137–158. https://doi.org/10.1016/B978-075063760-2/50029-9.
39. R. Vinuesa, S.L. Brunton, Enhancing computational fluid dynamics with machine learning, *Nat. Comput. Sci.* 2 (2022) 358–366. https://doi.org/10.1038/s43588-022-00264-7.
40. S.L. Brunton, B.R. Noack, P. Koumoutsakos, Machine learning for fluid mechanics, *Annu. Rev. Fluid Mech.* 52 (2020) 477–508. https://doi.org/10.1146/annurev-fluid-010719-060214.
41. Y. Zhang, Y. Wu, Introducing machine learning models to response surface methodologies, in: P. Kayaroganam (ed.), *Response Surface Methodology in Engineering Science*, IntechOpen, 2021. https://doi.org/10.5772/intechopen.98191.
42. S. Worku Kidane, *Application of Response Surface Methodology in Food Process Modeling and Optimization*, IntechOpen, 2021. https://doi.org/10.5772/intechopen.100113.
43. T.W. Simpson, J.D. Poplinski, P.N. Koch, J.K. Allen, Metamodels for computer-based engineering design: Survey and recommendations, *Eng. Comput.* 17 (2001) 129–150. https://doi.org/10.1007/PL00007198.
44. M. Raissi, P. Perdikaris, G.E. Karniadakis, Physics-informed neural networks: A deep learning framework for solving forward and inverse problems involving nonlinear partial differential equations, *J. Comput. Phys.* 378 (2019) 686–707. https://doi.org/10.1016/j.jcp.2018.10.045.
45. J. Yu, L. Lu, X. Meng, G.E. Karniadakis, Gradient-enhanced physics-informed neural networks for forward and inverse PDE problems, *Comput. Methods Appl. Mech. Eng.* 393 (2022) 114823. https://doi.org/10.1016/j.cma.2022.114823.
46. M. Lino, S. Fotiadis, A.A. Bharath, C.D. Cantwell, Current and emerging deep-learning methods for the simulation of fluid dynamics, *Proc. R. Soc. A* 479 (2023). https://doi.org/10.1098/rspa.2023.0058.
47. S. Cuomo, V.S. Di Cola, F. Giampaolo, G. Rozza, M. Raissi, F. Piccialli, Scientific machine learning through physics–informed neural networks: Where we are and what's next, *J. Sci. Comput.* 92 (2022) 88. https://doi.org/10.1007/s10915-022-01939-z.
48. Y. Jiang, E. Byrne, J. Glassey, X. Chen, Reduced-order modeling of solid-liquid mixing in a stirred tank using data-driven singular value decomposition, *Chem. Eng. Res. Des.* 196 (2023) 40–51. https://doi.org/10.1016/j.cherd.2023.06.019.
49. A. Mukhtar, A.S.H.M. Yasir, M.F.M. Nasir, A machine learning-based comparative analysis of surrogate models for design optimisation in computational fluid dynamics, *Heliyon* 9 (2023) e18674. https://doi.org/10.1016/j.heliyon.2023.e18674.
50. M. Golzarijalal, F. Zokaee Ashtiani, B. Dabir, Modeling of microalgal shear induced flocculation and sedimentation using a coupled CFD-population balance approach, *Biotechnol. Prog.* 34 (2018) 160–174. https://doi.org/10.1002/btpr.2580.

Numerical Modelling of the Hydrodynamics and Fluid–Particle Interactions in a Rotatory Vortex Pelleting (RVP) Reactor

6

Benjamin Oyegbile

6.1 INTRODUCTION

Efficient fluid–particle mixing and dispersion are highly indispensable in many industrial, pharmaceutical, biological, and environmental processes involving fluid–particle interactions. In fact, mixing, transport, aggregation, and deposition of particulate solids occur simultaneously in many natural, biological, and engineered processes [1–5]. Therefore, in studying the dynamics of these systems, the spatial and temporal evolution of the discrete phase properties dispersed within the continuous phase is of great interest. In solid–liquid mixing especially in agitated vessels, there exist three distinct phases of particle dispersion within the fluid phase, namely on-bottom, off-bottom, and fully suspended phases [6]. The mixing efficiency in any fluid–particle system largely depends on the dispersion uniformity in terms of the spatial and temporal distribution and homogeneity of the discrete phase within the continuous phase. This can be assessed in terms of specie attributes such as suspended speed, solids mass concentration, volume fraction, number density, and size distribution within the mixing vessel [7].

Computational fluid dynamics (CFD) offers a flexible and robust platform for the investigation of fluid–particle interactions in simple to complex reactor systems and a multitude of flow scenarios occurring in a variety of process conditions [7]. The application of CFD to model particulate flow problems has been growing in popularity over the past decade. Several modelling techniques have been proposed with some encouraging results in terms of both model predictions and computational demand. The Eulerian–Eulerian model, Eulerian–Lagrangian models such as the two-fluid model (TFM), dense discrete phase

DOI: 10.1201/9781003518426-6

model (DDPM), discrete element method (DEM), kinetic theory of granular flow (KTGF), frictional stress model (FSM), etc., are a few of the widely used modelling approaches in the scientific community especially in the modelling of dense particle-laden flows [8, 9].

In the numerical simulation of the hydrodynamics of the discrete phase dispersed within a continuous phase, an accurate resolution of the fluid–particle, particle–particle, and particle–wall interaction forces is highly indispensable to obtain an accurate result. The ability to predict the dynamics of the dispersed phase, i.e., dispersion, transport, coalescence/agglomeration, breakage, and deposition, depends on an accurate resolution of all the forces experienced by individual particles within a turbulent flow field. However, it is worth pointing out that quantifying all the forces that mediate fluid–particle interactions is a daunting task [8]. The particle–particle and particle–wall interaction forces can be implicitly resolved using the kinetic theory of granular flow (KTGF) or explicitly resolved using the discrete element method (DEM). This is then coupled to the fluid flow via mass, momentum, and energy exchange between the continuous and the dispersed phase to account for the fluid–particle interactions. Coupled CFD–DEM is widely used for the simulation of many flow problems involving fluid–particle interactions due to its high degree of accuracy in resolving nearly all the fluid–particle and particle–particle interaction forces [1, 10–14].

In spite of its accuracy, one main drawback of the DEM approach is the huge computational demand even for simple multiphase flows with low to moderate solids concentration, which limits the scope of its application [9, 15]. Over the past decade, there has been a steady improvement in computing capacity and coupled CFD–DEM is increasingly being used to simulate flows with hundreds of thousands to millions of particles by using efficient numerical algorithms and parallelisation [10]. However, in most of the practical industrial flows, the number of particles can range from several millions to billions, which significantly increases the simulation time, thereby limiting the scope of applications of DEM on an industrial scale [11, 15]. Coupled DPM–KTGF approach, on the other hand, offers a more reasonable compromise in terms of the computational demand and numerical accuracy and therefore more computationally efficient [12, 13]. Several authors have argued that highly coupled simulations are sometimes unnecessary, especially for dilute flows with binary collisions, as reasonable abstractions can be made while still capturing the essential features of the flow [10, 14, 16]. The KTGF model describes an interaction force $F_{\text{interaction}}$ model for a particulate flow based on the particulate phase solids stress tensor τ_s using the kinetic theory of granular flow. The model, which is analogous to the kinetic theory of gas that describes the thermal motion of molecules in a dense gas, is used to derive the solids pressure and solids stresses arising from the random motion of the discrete particles, taking into account the inelastic nature of the granular phase. This coupled model also takes into account the fact that when particles collide, energy is dissipated in the form of deformation and heat by using coefficients of elastic restitution to characterise the loss of momentum during collisions. This approach is ideally suitable for dilute flows, but may be extended to prolonged and friction-dominated particle–particle interactions in dense flows by using the frictional stress (FSM), and dense discrete phase model (DDPM). The idealized particle–particle interaction forces are not explicitly resolved as in the soft sphere or classical DEM approach but are modelled, thereby resulting in a much faster computational turnaround, especially for simple fluid–particle interactions [10]. Several authors have also successfully employed this alternative modelling approach to reduce the computational demand by applying KTGF approach to model the particle–particle interactions based on the continuum approach within the Eulerian framework [17, 18–20].

A quick survey of the literature reveals a large body of scientific work on the applications of DEM as compared to the coupled DPM–KTGF approach. Cloete et al. applied different fluid particle models to study dilute gas–solid flow using two modelling approaches, namely Eulerian–Eulerian two-fluid model (TFM–KTGF) and Eulerian–Lagrangian (DDPM–KTGF) model. These models were validated against each other and the results show a close agreement between the models, with the observed minor differences in the results attributed to the formation of delta shocks which was better captured by the DDPM–KTGF approach [15]. In a similar study by Gannt and Gatzke, coupled CFD–DEM and DPM–KTGF models were simultaneously applied to the simulation of particulate flow in a high-shear mixer [21]. The result from their work shows a good agreement between the two fluid–particle models under idealized conditions, while variations were observed as the process conditions become less ideal. The observed deviation was attributed to some of the assumptions regarding particle sphericity, rigidity, and smoothness in the KTGF model in resolving the particle–particle interactions [11]. In addition to incorporating the Brownian and shear forces in the KTGF model, Weber and Hrenya [22] in their study extended the scope of KTGF approach to include non-contact cohesive particle forces such as the van der Waals forces using the square well model. This study opened up the possibility of this model being used to account for particle–particle collisions, coalescence, and aggregation due to non-contact forces.

Therefore, the aim of this chapter is to demonstrate how to carry out a qualitative and quantitative analysis of the hydrodynamics of granular particle dispersion using a rotatory vortex pelleting (RVP) reactor by employing a Eulerian–Lagrangian formulation with one-way coupling between the discrete phase and the continuous phase. Owing to the complexity of the agglomeration process, modelling effort is devoted to the investigation of the particle–particle and particle–wall interactions in the reactor. The numerical simulation was performed by coupling the discrete phase model (DPM) with the KTGF model to study the spatiotemporal properties of the dilute particulate flow as well as the particle trajectory using ballotini and sand as model particles. The model was subsequently used to study the influence of the operating conditions such as the restitution coefficients, disc rotation speed, and particle dispersity on the flow characteristics. The modelling results were validated using the experimental data from particle image velocimetry (PIV) analysis of the single-phase flow. The CFD model can also be extended to model dense particulate flows by extending the capability of both the KTGF and DPM models by incorporating the frictional stress model (FSM) in combination with dense discrete particle model (DDPM) in order to account for the increased particle–particle contacts and frictional effects.

6.2 MODEL DEVELOPMENT

6.2.1 Continuous Phase Conservation Equations

The dispersion of particles in a shear flow is modelled using a Eulerian–Lagrangian framework with the particle–particle interaction force $F_{\text{interaction}}$ accounted for using the kinetic theory of granular flow. The Reynolds-averaged Navier–Stokes (RANS) conservation equations of mass and momentum for an incompressible flow can be written as follows (Equations 6.1–6.2) [17, 23]:

$$\frac{\partial \rho}{\partial t} + \nabla \cdot (\rho \vec{v}) = S_{c,d_p} \tag{6.1}$$

$$\frac{\partial}{\partial t}(\rho \vec{v}) + \nabla \cdot (\rho \vec{v}\vec{v}) = -\nabla \vec{p} + \nabla \cdot \mu_{eff} \left[\nabla \vec{v} + (\nabla \vec{v})^T \right] + \rho \vec{g} + \vec{S}_{m,d_p} \tag{6.2}$$

where ρ is the density, p is the static pressure, $\vec{v}$ is the velocity vector, S_{c,d_p} is the source term that represent the mass added to the continuous phase by the dispersed phase or any user-defined mass source and this is zero by default, $\vec{g}$ is the gravitational force, and $\vec{S}_{m,d_p}$ is a source term representing the interphase exchange of momentum arising from particle–particle and fluid–particle interactions. The effective viscosity μ_{eff} and the momentum transfer between the discrete and the continuous phase are given in Equations (6.3) and (6.4) [17]:

$$\mu_{eff} = \mu_m + \mu_t = \mu_m + \rho C_\mu \frac{k^2}{\varepsilon} \tag{6.3}$$

$$\vec{S}_{m,d_p} = \Sigma \left(\frac{18 \mu C_D R_e}{\rho_p d_p{}^2 24} (u_p - u) + F_{other} \right) \grave{U} m_p \, \Delta t \tag{6.4}$$

6.2.2 Discrete Phase Force–Balance Equations

The force balance equation obtained from Newton's second law that governs the particle acceleration along with the particle translational equation is used to resolve the particle trajectory. This is achieved by stepwise integration of the trajectory equation (Equation 6.5) along with any auxiliary equations describing any heat and mass transfer from the discrete phase to the continuous phase, while the particle rotational velocity is obtained by solving an additional ordinary differential equation for particle's angular momentum (Equation 6.6) [6, 17]:

$$m\frac{d\vec{v}}{dt} = \overrightarrow{F_D} + \overrightarrow{F_P} + \overrightarrow{F_{VM}} + \overrightarrow{F_G} + \overrightarrow{F_B} + \overrightarrow{F_L} + \overrightarrow{F_{Int}} \tag{6.5}$$

$$I_p\frac{d\vec{\omega}_p}{dt} = \frac{\rho_f}{2}\left(\frac{d_p}{2}\right)^5 C_\omega \vec{\Omega} = \vec{T} \tag{6.6}$$

where F_D, F_P, F_{VM}, F_G, F_L, and F_{int} represent the drag, pressure gradient, virtual mass, gravitational, lift, and the granular interaction forces, respectively. However, the gravitational, buoyant, and the Basset history forces were neglected in this study due to the rotational nature of the flow while the pressure gradient and virtual mass forces were considered to be important as the ratio of the phase densities is greater than 0.1 [17]. Other non-contact interfacial forces that mediate particle–particle interactions such as van der Waals attraction, Born repulsion, electric double layer (EDL), as well as the Basset history and viscous stress forces due to fluid–particle interactions were neglected in this study to lower the computational cost. Table 6.1 provides a summary of CFD modelling approaches and the treatment of dominant particle–particle and fluid–particle forces.

TABLE 6.1 Multiphase Models and Fluid–Particle Interaction Forces

CFD MODEL	*CONTACT P–P INTERACTION FORCES*	*NON-CONTACT DLVO P–P FORCES*	*F–P FORCES*
PBM	None	Electrostatic, van der Waals, bridging	Drag, rotational drag, lift, rotational lift, pressure gradient, virtual mass, Brownian, viscous, etc.
DPM	None	None	Drag, rotational drag, lift, rotational lift, pressure gradient, virtual mass, Brownian, viscous, hydraulic retardation, etc.
DPM–square well	Modelled	Electrostatic, van der Waals	Drag, rotational drag, lift, rotational lift, pressure gradient, virtual mass, Brownian, viscous, hydraulic retardation, etc.
Euler–granular	Modelled	None	Drag, rotational drag, lift, rotational lift, pressure gradient, virtual mass, Brownian, viscous, hydraulic retardation, etc.
DDPM–KTGF	Modelled	None	Drag, rotational drag, lift, rotational lift, pressure gradient, virtual mass, Brownian, viscous, hydraulic retardation, etc.
DEM	Resolved	Electrostatic, van der Waals, bridging, hydration, hydrophobic	Drag, rotational drag, lift, rotational lift, pressure gradient, virtual mass, Brownian, viscous, hydraulic retardation, etc.

Note: P–P = Particle–particle forces and F–P = Fluid–particle forces.

6.2.3 Determination of the Granular Collisional Forces

The granular collision or discrete phase interaction force F_{int} in Equation (6.7) is modelled from a correlation derived from the solids phase stress tensor due to the random collisions and translations as a result of fluctuating particle velocities using the kinetic theory of granular flow in Equation (6.8) [13, 15, 24]:

$$F_{\text{int}} = -\frac{1}{\rho_s}\nabla \cdot \vec{\tau_s} \tag{6.7}$$

where ρ_s represents the solids density, and $\vec{\tau_s}$ is the solids stress tensor which includes the solids pressure, shear stress, and normal stress [15]. The solids stress tensor is given by Equation (6.8) as follows:

$$\vec{\tau_s} = -p_s\bar{I} + \alpha_s\mu_s\left(\nabla\overline{v_s} + \left(\nabla\overline{v_s}\right)^T\right) + \alpha_s\left(\lambda_s - \frac{2}{3}\mu_s\right)\nabla\cdot\overline{v_s}\bar{I} \tag{6.8}$$

where p_s, μ_s, and λ_s are the solids pressure, shear viscosity, and bulk viscosity, respectively. The solids pressure which is a measure of the momentum transfer due to the fluctuating motion of particles and their collisions is given by Equation (6.9):

$$p_s = \alpha_s\rho_s\Theta_s + 2\rho_s\left(1+e_{ss}\right)\alpha_s^{\,2}g_{0,ss}\Theta_s \tag{6.9}$$

The equation mentioned above describes a formulation given by Lun et al. [17] in which the first term on the right-hand side represents the kinetic contribution with the second term representing the collisional contribution, while the frictional contribution can be neglected for particle spacing below the packing limit [17]. The term e_{ss} is the coefficient of restitution for particle–particle collisions, $g_{0,ss}$ is the radial distribution function which is a correlation factor that modifies the probability of collisions when the solid granular phase is close to the packing limit (Equation 6.10), while Θ_s is the solids granular temperature which is a measure of the random motion of the particles and it is proportional to the kinetic energy of the fluctuating particle motion. It is defined as one-third of the square of the fluctuating particle velocities [25] (Equation 6.11). The granular temperature is obtained by solving the algebraic form of the transport equations of the kinetic theory (Equation 6.12), in which the convection and diffusion terms in the transport equations are neglected [21, 23].

$$g_{0,ss} = \left[1 - \left(\frac{\alpha_s}{\alpha_{s,\max}}\right)^{\frac{1}{3}}\right]^{-1} \tag{6.10}$$

$$\Theta_s = \frac{1}{3}\left(u_{s,i}u_{s,i}\right) \tag{6.11}$$

$$\frac{3}{2}\left[\frac{\partial}{\partial t}\left(\rho_s\alpha_s\Theta_s\right) + \nabla\cdot\left(\rho_s\alpha_s\vec{v_s}\Theta_s\right)\right] = \left(-p_s\bar{I} + \overline{\overline{\tau_s}}\right)6\nabla\vec{v_s} + \nabla\cdot\left(k_{\Theta s}\nabla\Theta_s - \gamma_{\Theta s} + \varphi_{ls}\right) \tag{6.12}$$

The solids stress which is composed of the shear viscosity due to the kinetic motion $\mu_{s,kin}$, collisional interactions $\mu_{s,col}$, and frictional particle interactions $\mu_{s,fr}$ (Lun et al. and Gidaspow et al. [17]) given in Equations (6.13) and (6.14) and the bulk viscosity which account for the particle resistance to expansion and compression (Lun et al. [17]) given by Equation (6.15) [17]. The frictional contribution to the shear viscosity can be neglected in Equation (6.14) for dilute flows below the packing limit:

$$\mu_s = \mu_{s,col} + \mu_{s,kin} + \mu_{s,fr} \tag{6.13}$$

$$\mu_s = \frac{4}{5}\alpha_s^{\ 2}\rho_s d_s g_{0,ss}\left(1+e_{ss}\right)\left(\frac{\Theta_s}{\pi}\right)^{\frac{1}{2}} + \frac{10\rho_s d_s\sqrt{\Theta_s\pi}}{96\alpha_s\left(1+e_{ss}\right)g_{0,ss}}\left[1+\frac{4}{5}g_{0,ss}\alpha_s\left(1+e_{ss}\right)\right]^2 \tag{6.14}$$

$$\lambda_s = \frac{4}{3}\alpha_s\rho_s d_s g_{0,ss}\left(1+e_{ss}\right)\left(\frac{\Theta_s}{\pi}\right)^{\frac{1}{2}} \tag{6.15}$$

6.2.4 Turbulent Dispersion of the Granular Particles

The effect of the turbulent continuous phase flow on the discrete phase dispersion can be determined using stochastic tracking or deterministic methods. The discrete random walk (DRW) formulation is available in ANSYS Fluent. This stochastic method based on the eddy lifetime approach incorporates the influence of turbulent velocity fluctuations on the particle trajectories by deriving the fluctuating velocity u' component of the instantaneous velocity u via a Gaussian distribution function where σ refers to the *rms* value of the fluctuating velocity u'. These derivations are given in Equations (6.16) and (6.17) where ζ is a vector formed by normally distributed random number, k is the kinetic energy, and σ is the root mean square value of the fluctuating velocity with the assumption of an isotropic turbulence [4]:

$$u' = \zeta\sigma \tag{6.16}$$

$$\sigma = \sqrt{\frac{2k}{3}} \tag{6.17}$$

6.2.5 Turbulence Modelling and Turbulence Modifications

The $k - \varepsilon$ dispersed turbulence model is used to account for the effect of turbulence on the multiphase flow by calculating the Reynolds stress tensor using an extended form of the single phase RNG-$k - \varepsilon$ two-equation model with curvature correction to improve the turbulence predictions due to the rotational nature of the flow [17]. Using a modified form of $k - \varepsilon$ model transport equations (Equations 6.18 and 6.19), the turbulence predictions are obtained for the continuous phase while turbulence quantities for the dispersed phase are given in terms of the mean characteristics of the continuous phase and the ratio of the particle relaxation time and the eddy-particle interaction time from Tchen theory [18, 19]. The effect of increasing turbulence (Reynolds number and eddy sizes) on the drag predictions was accounted for by using a drag modification factor proposed by Brucato et al. [18], which expresses drag coefficient as a function of the ratio of particle diameter and Kolmogorov length scale (Equation 6.20) [18]. However, the effect of the Magnus rotational lift force of the fluid was not considered in coupled DPM calculations due to the limitations in the numerical code. Therefore, particle rotation was only considered from the torque on the particle arising particle–wall interactions in the coupled calculations [24]. Turbulence damping, modulation, or attenuation by the suspended particles especially at high solids loadings may also be incorporated into the model by implementing a two-way turbulence coupling in the numerical calculations using the formulation proposed by Amsden, O'Rourke, and Faeth [19]:

$$\frac{\partial}{\partial t}\left(\alpha\rho k\right) + \nabla\cdot\left(\alpha\rho\overrightarrow{U}k\right) = \nabla\cdot\left(\alpha\left(\mu + \frac{\mu_t}{\sigma_k}\right)\nabla k\right) + \alpha G - \alpha\rho\int + \alpha\rho\Pi \tag{6.18}$$

and

$$\frac{\partial}{\partial t}(\alpha\rho k)+\nabla\cdot\left(\alpha\rho\vec{U}\varepsilon\right)=\nabla\cdot\left(\alpha\left(\mu+\frac{\mu_t}{\sigma_\varepsilon}\right)\nabla\varepsilon\right)+\alpha\frac{\varepsilon}{k}\left(C_{1\varepsilon}G-C_{2\varepsilon}\rho\varepsilon\right)+\alpha\rho\Pi \tag{6.19}$$

$$\eta=1+\left(\frac{C_D-C_{D0}}{C_{D0}}\right)=K\left(\frac{d_p}{\lambda}\right)^3 \tag{6.20}$$

where C_D is the computed drag coefficient due to the turbulence effect, C_{D0} is the drag coefficient without turbulence modification, K is a constant with value of 6.5×10^{-6}, d_p is the particle diameter, and λ is the Kolmogorov length scale [17, 18]. In addition to the stresses predicted by the kinetic theory, the KTGF model may be extended by include the frictional effect of the granular phase when the solids volume fraction exceeds the critical value using the frictional stress model (FSM) which incorporate frictional viscosity and frictional pressure terms in Equations (6.9) and (6.14) [17]. The DDPM form of Equations (6.1) and (6.2) is used when the dispersed phase is close to the packing limit, especially in the case of dense particulate flows. These two additional equations are however omitted from this investigation due to the dilute nature of the flow considered. However, these parameters may be included depending on the flow scenario being modelled [17].

6.3 IMPLEMENTATION OF THE NUMERICAL SOLUTION

6.3.1 Flow Domain and Geometry

A bench-scale rotor-stator vortex reactor for the agglomeration of dilute particulate suspension which was developed in-house was employed in carrying out this study. Detailed information on the reactor setup and the process conditions can be found in our earlier publications [20, 26]. The geometric design of the reactor and the subsequent mesh generation of the flow domain are briefly presented. In implementing a numerical scheme to solve the above set of equations describing the particle–particle, particle–wall, and fluid–particle interactions within the reactor, a solid model of the reactor depicted schematically in Figure 6.1 was first created using commercial solid modelling software (ANSYS DesignModeler 2024R2, ANSYS Inc.).

6.3.2 Flow Description and PIV Measurements

The flow within the reactor is an enclosed Batchelor flow, which is an axis-symmetrical forced vortex flow with separated boundary layers at the rotor and the stator and a radially pumping effect towards the stator. Particle dispersion within the reactor is largely shear driven with patches of turbulence especially above the critical disc or local rotational Reynolds number. Batchelor proposed a model to describe the fluid flow structure in an enclosed rotor-stator disc system with the fluid being pumped axially outwards from the rotor boundary layer [27]. Characterisation of the flow dynamics was carried out via a particle image velocimetry (PIV) analysis. The details of the PIV flow measurements are available elsewhere [20]. For the sake of brevity, the discussion will focus on the description of the flow tracing particles which consists of a mixture of hollow glass spheres and silver-coated glass spheres in ratio 1:1 with a median diameter of 10 μm supplied Dantec Dynamics A/S, Denmark. The particles were selected due to their good scattering efficiency and sufficiently small velocity lag [28].

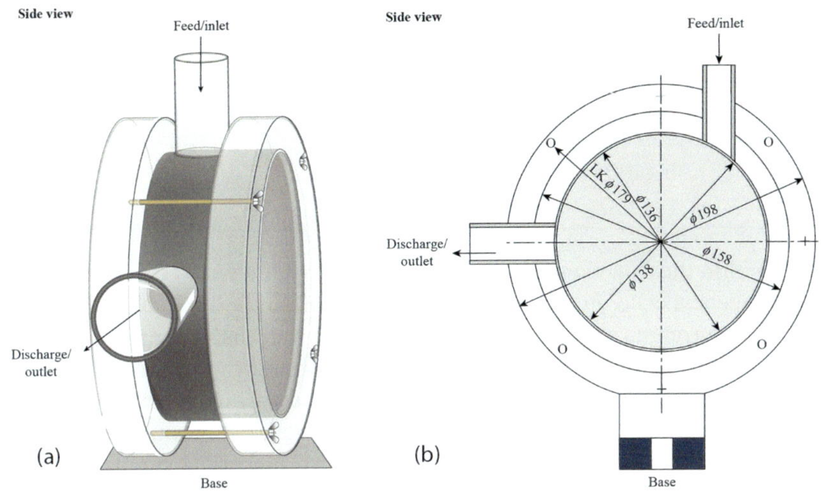

FIGURE 6.1 Schematic illustration of the 3D CAD model of the flow geometry (not to scale, all dimensions in millimetres: (a) front view and (b) side view.

6.3.3 Solution Control and Mesh Convergence Analysis

In order to evaluate the effect of the mesh density on the solution accuracy, four different mesh sizes were generated using the ANSYS Mesher, after which a sensitivity analysis was performed to determine an optimal mesh size. Thereafter, a high-quality hexahedral mesh of approximately 431,000 cells (mesh 2) was selected for the numerical simulation based on the sensitivity analysis shown in Figure 6.2. Table 6.2 gives a quantitative measure of the mesh quality for all the grid sizes considered. Monodispersed sand and polydispersed ballotini particles were selected as model granular materials for the particle tracking calculations. They were selected because they give a realistic representation of real-world particulate systems. The mesh sensitivity study was conducted by carrying out the numerical simulation on successively fine grids using four different mesh resolutions. Thereafter, the results of the tangential velocities

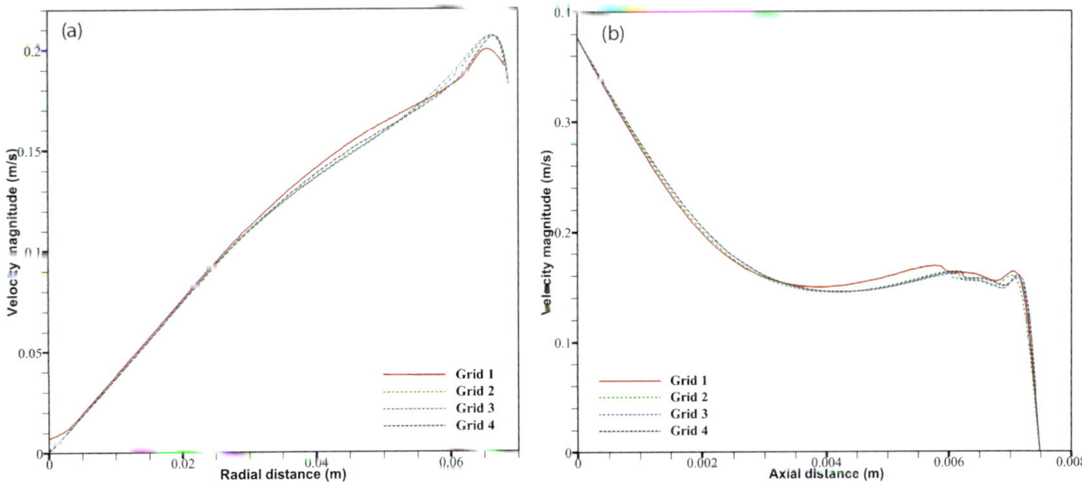

FIGURE 6.2 Flow velocity profiles of different grid sizes on the *YZ* and *XZ* planes for RNG $k - \varepsilon$ model $N =$ 70 rpm: (a) $x = 0.003725$ and (b) $y = 0.045$.

TABLE 6.2 Mesh Metrics for the Structured Hexahedral Grid

MESH NUMBER	*NODES*	*SKEWNESS (AVERAGE)*	*ORTHOGONAL QUALITY (AVERAGE)*
Grid 1	51,448	9.30×10^{-2}	0.9893
Grid 2	499,848	5.82×10^{-2}	0.9923
Grid 3	995,180	4.78×10^{-2}	0.9941
Grid 4	1,434,632	4.83×10^{-2}	0.9924

on selected planes within the flow domain along the axis of rotation were compared for all the grids to establish their grid independence. Finally, a mesh resolution of 431,000 hexahedral cells was chosen to run the simulation to a converged solution based on the fact that the relative errors between the mesh sizes were quite minimal (below 1% difference). This optimized mesh was thereafter employed for subsequent computations of all flow properties. The residuals, moment, and average kinetic energy are used as the convergence parameters for the numerical solution, while the solution is considered to have converged when the residual reduces to an acceptable level of 1×10^{-3}. All numerical simulations were performed using laptop with a quad-core i5-6500, 3.20 GHz 6MB cache CPU and 32 GB RAM. Figure 6.2 shows the vertical and horizontal tangential velocity profiles for all the four grids employed for the sensitivity analysis. The profiles were drawn along the *YZ* and *XZ* planes at $x = 0.003725$ and $y = 0.045$, respectively, for a rotation speed of 70 rpm.

6.3.4 Numerical Solution Strategy and Boundary Conditions

The mathematical model developed in Section 6.2 was solved numerically on a computational grid to predict the fluid–particle interactions in a bench-scale vortex reactor using point-particle method where moving particles are treated as mass points. In this so-called hard-sphere or event-driven approach, some abstractions are made regarding the particle shape, volume, and fluid flow around the particles [4, 29]. The turbulent fluid flow was modelled by solving incompressible Reynolds-averaged Navier–Stokes (RANS) equations with scalable wall functions, while the discrete phase was modelled using coupled DPM–KTGF Lagrangian particle tracking equations, both of which were implemented in a commercial CFD software package (ANSYS Fluent v18.2, ANSYS Inc.). A converged steady-state fluid flow field was weakly coupled to a transient DPM particle tracking model via a one-way coupling to account for the fluid–particle interactions and specie distribution within the flow domain. This was achieved by increasing the number of continuous phase calculations between the trajectory calculations after a converged steady-state solution has been obtained. A one-way coupling is justified in the present study considering that the mass loadings and volume fractions are quite low at less than 10% and 10^{-4}, respectively, and the mean particle sizes are much less than the Kolmogorov length scale $\approx$ 12 μm [4, 24, 30]. This implies that the discrete phase is primarily influenced by the hydrodynamics of the continuous phase.

The calculation of the particle–particle contact forces was not explicitly resolved but modelled via KTGF. However, this event-driven approach takes into account the energy dissipation and resultant loss of momentum during collisions through the use of restitution coefficients. The Gidaspow model was employed to compute the collisional and kinetic viscosities while a formulation by Lun et al. [17] was used for the solids pressure (Equations 6.9, 6.13, and 6.14, respectively) [25]. Calculations of the particle trajectories were done via a stepwise integration of the trajectory equations from the converged CFD velocity field using an implicit integration scheme [21, 29]. In these idealised particle–particle interactions, particles were injected from surfaces within the reactor wheel space in the direction of the flow. A reflective-type discrete phase boundary condition was imposed on all the walls with no-slip boundary condition for the continuous phase [31]. Realistic values of the normal particle restitution and wall reflection coefficients were chosen for the normal forces, while the tangential linear velocity after collision or reflection is a function of the linear velocity and the angular velocity due to particle rotation. However, in

TABLE 6.3 Thermophysical Properties of Fluid and Solid Particles and Simulation Parameters

	MATERIALS		
PROPERTIES	*WATER*	*MONO-SAND*	*POLY-BALLOTINI*
Fluid density (kgm^{-3})	998		
Dynamic viscosity (kgm^{-1}s^{-1})	8.94×10^{-4}	—	—
Temperature (°C)	25		
Maximum particle size (μm)	—	—	20
Mean particle sizes (μm)	—	2, 6, 10	2, 6, 10
Minimum particle size (μm)	—	—	1
Particle density (kgm^{-3})	—	2650	2500
Particle shape	—	Spherical	Spherical
Number of injected particles (—)	—	3.6×10^6, 7.2×10^6	1.25×10^7, 2.5×10^7
Solids volume fractions (—)	—	8.2×10^{-6}, 1.6×10^{-5}	8.7×10^{-6}, 1.7×10^{-5}
Particle injection type	—	Surface	Surface
Total injection time (s)	—	0.005, 0.01	0.005, 0.01
Phase coupling scheme	—	One way	One way
Mixing speeds (rpm)	—	70–130	70–130
Particle restitution coefficients (—)	—	0.20–0.99	0.20–0.99
Wall reflection coefficients (—)	—	1	1
Friction coefficient (—)	—	0.2	0.2
Particle tracking time step Δt (s)	—	1×10^{-4}	1×10^{-4}

situations where the particle rotation is not enabled or compatible with the trajectory calculations such as in the case of moving reference frame, a formulation by Tsuji et al. was employed to determine the values of tangential or sliding force with default values used for the friction coefficients [17]. The default values of the friction coefficient as well as the restitution coefficients specified in Table 6.3 ensure that the particles retain much of their momentum in the normal and tangential direction after collision. The angular momentum in the wall normal direction is not affected by collision which results in an unchanging particle angular velocity in the case of adhesive impact considered in this study [17]. The coupled algorithm was used for the pressure–velocity coupling while the QUICK scheme was applied to the rest of the flow variables. Table 6.3 shows a summary of the phase properties and the particle injection schemes for the DPM calculations.

6.3.5 Theoretical and Experimental Validation of CFD Model

The validation of the numerical model from this study was done based on the theoretical analysis of the rotor-stator flow or the so-called Batchelor model predictions as well as the PIV flow measurements for the single-phase flow (Figure 6.3a, b) [20, 27]. Figure 6.3a shows the Batchelor and the CFPD models' horizontal velocity profiles with two identical separated boundary layers on the rotor and the stator and a middle fluid rotating core. However, the gap ratio used in the Batchelor measurements was quite different from the one used in the present study which might be responsible for the variations observed in the fluid rotating core [32]. Figure 6.3b shows the vertical velocity and vorticity profiles, which gives a reasonably good agreement between the CFPD model predictions and PIV measurements. Owing to some of the limitations reported in the PIV measurements [20], it could be argued that the flow parameters were under-predicted by the PIV analysis. Therefore, the observed deviations were mainly due to some of the experimental limitations rather than the robustness of the CFD model or the numerical code.

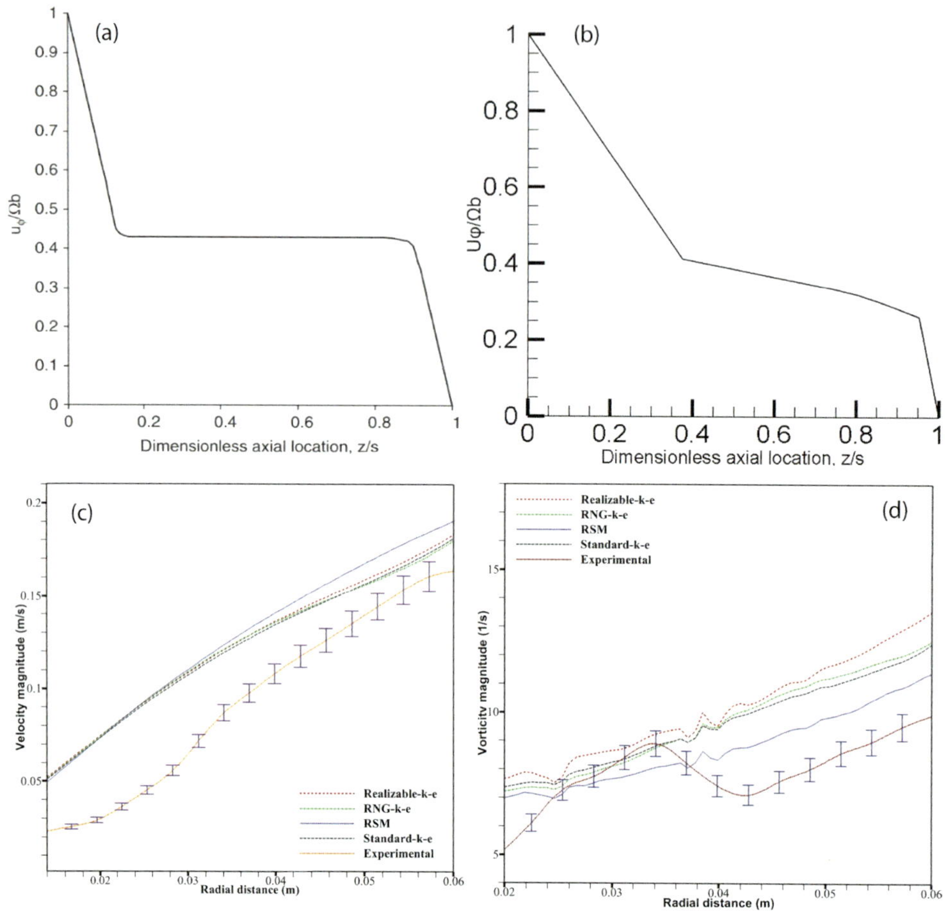

FIGURE 6.3 Horizontal and vertical velocity and vorticity profiles on the *YZ* plane at $x = 0.003725$ and the *XZ* plane at $y = 0.068$: (a) horizontal velocity profiles for CFD model, (b) horizontal velocity profile for Batchelor model, (c) vertical velocity profiles for CFD and PIV analysis, and (d) vertical vorticity profiles for CFD and PIV analysis.

6.4 CFD RESULTS AND DISCUSSION

6.4.1 Evolution of the Discrete Phase Concentration

The spatiotemporal evolution of the flow properties in terms of the discrete phase solids concentration for the polydispersed ballotini particles is presented in Figures 6.4 and 6.5. It is observed from the figures that there is an accumulation of higher particle concentration near the rotating shaft as in conventional stirred tank at a vertical plane along the impeller [11]. This trend is seen to be consistent and increases throughout the duration of the particle tracking. The area around the shaft does not seem to promote particle mixing as a significant particle holdup was observed in this region due to low turbulence intensity. One way to improve particle mixing in this region is by incorporating baffles or blades into the design of the rotating shaft to promote mixing in this region.

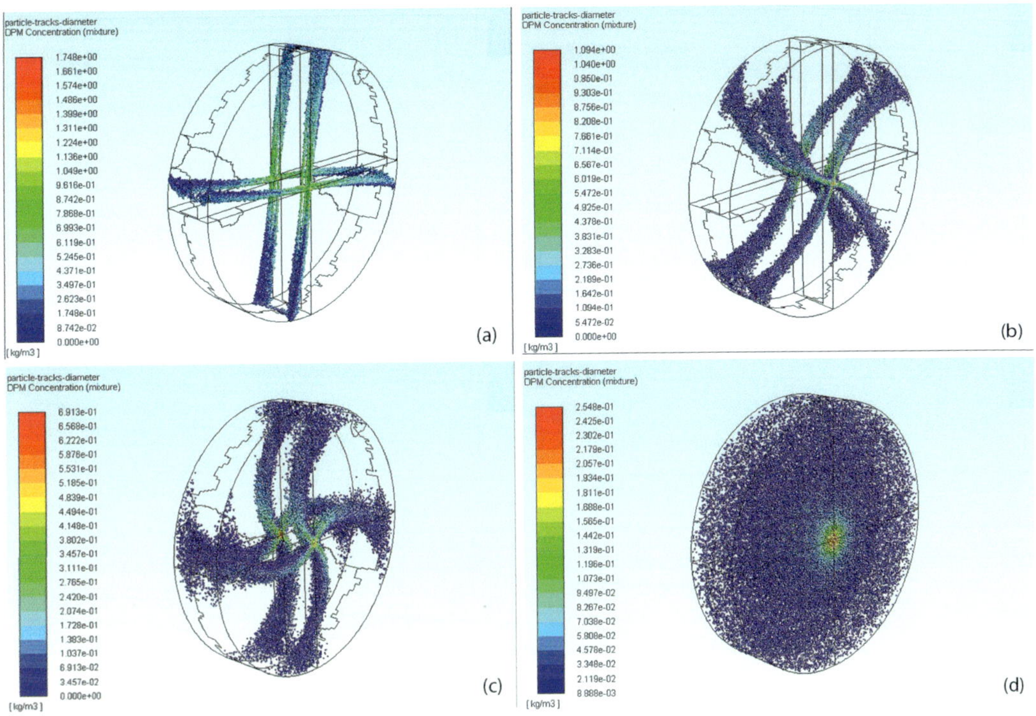

FIGURE 6.4 Spatiotemporal evolution of the instantaneous solids concentration of ballotini particles N = 70 rpm, $e_{ss} = 0.99$, $d_{50,\text{ballotini}} = 10\ \mu\text{m}$: (a) $t = 0.01$ s, (b) $t = 0.05$ s, (c) $t = 0.1$ s, and (d) $t = 1$ s.

6.4.2 Evolution of the Discrete Phase Size Distribution

The spatiotemporal evolution of the particle size distribution and dispersity within the reactor is presented in Figures 6.6 and 6.7 which clearly show a sustained change in the particle dispersity with the tracking time. A nearly uniform particle dispersity with somewhat narrow distribution is observed after 1 s due to the onset of steady-state conditions after this period. Steady-state conditions can also be achieved due to an equilibrium of particle collisions, coalescence, growth, breakage, and re-agglomeration [1, 11]. However, this was not investigated in this study. In terms of the spatial distribution, there is a wider particle size distribution axially towards the reactor stator (Figure 6.7b, c) while there appears to be a fairly uniform radial distribution of the particle sizes across the reactor diameter (Figure 6.7a). The wider particle size distribution observed axially towards the stator might be attributed to the pumping effect of the Batchelor flow in this direction [32].

6.4.3 Effect of Particle Size and Dispersity on the Bulk Viscosity

In order to understand the influence of the particle sizes and particle dispersity on the granular properties, numerical simulations were carried out with varying particle sizes starting with a converged solution for the largest particle size to study the effect of changing particle sizes on the granular bulk viscosity. Figures 6.8–6.11 show the solids bulk viscosity contours and profiles computed at three points along the reactor cavity, namely near the rotor, at the centre, and near the stator for monodispersed sand and polydispersed ballotini particles, respectively. It is observed that the bulk viscosity, which is a function

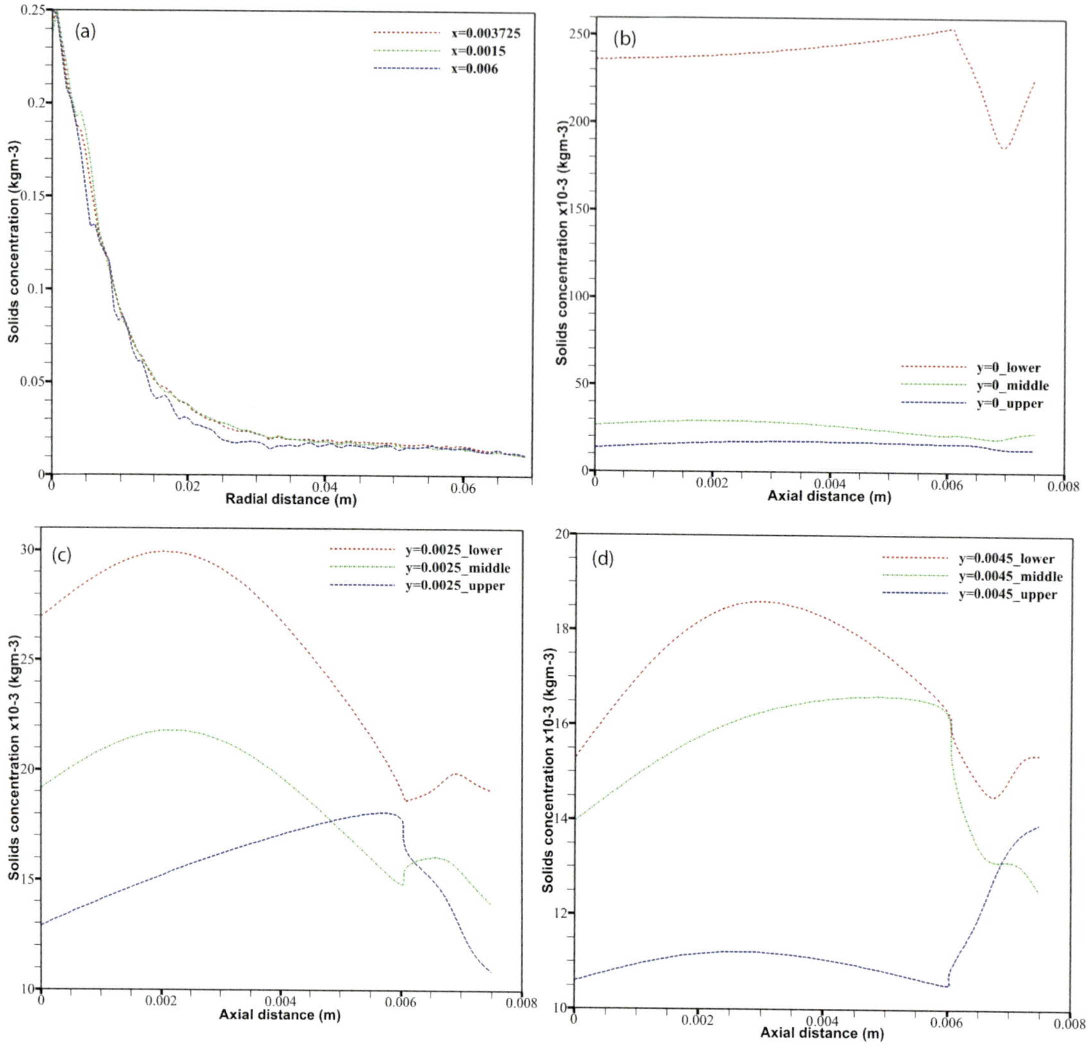

FIGURE 6.5 Vertical and horizontal profiles of the instantaneous solids concentration of ballotini particles on the *YZ* and *XZ* planes after 1 s N = 70 rpm, e_{ss} = 0.99, $d_{50,ballotini}$ = 10 μm: (a) vertical profiles, (b) horizontal profile y = 0, (c) horizontal profile y = 0.025, and (d) horizontal profile y = 0.045.

of particle diameter (Equation 6.12), increases with the particle sizes and this trend is consistent across the cavity (Figure 6.8a–c). It can be seen that the highest values of the bulk viscosities are obtained near the rotor due to high particle velocities and concentration at this position as compared to the centre of the cavity and near the stator. The increase in granular bulk viscosities due to the increasing diameter was due to the increasing resistance of the particles to compression and expansion as the surface area increases. This is consistent with similar studies elsewhere and this implies that the changing particle sizes influences the hydrodynamics of the reactor system [25]. A similar trend is observed in the case of ballotini particles (Figures 6.10 and 6.11) which seems to imply that particle dispersity does not seem to have a strong influence on the granular bulk viscosities. Therefore, it can be argued that the granular bulk viscosity is a function of the particle diameters rather than the dispersity of the particles. However, ballotini particles exhibit a lower bulk viscosity when compared to the sand particles which might be attributed to their smaller densities.

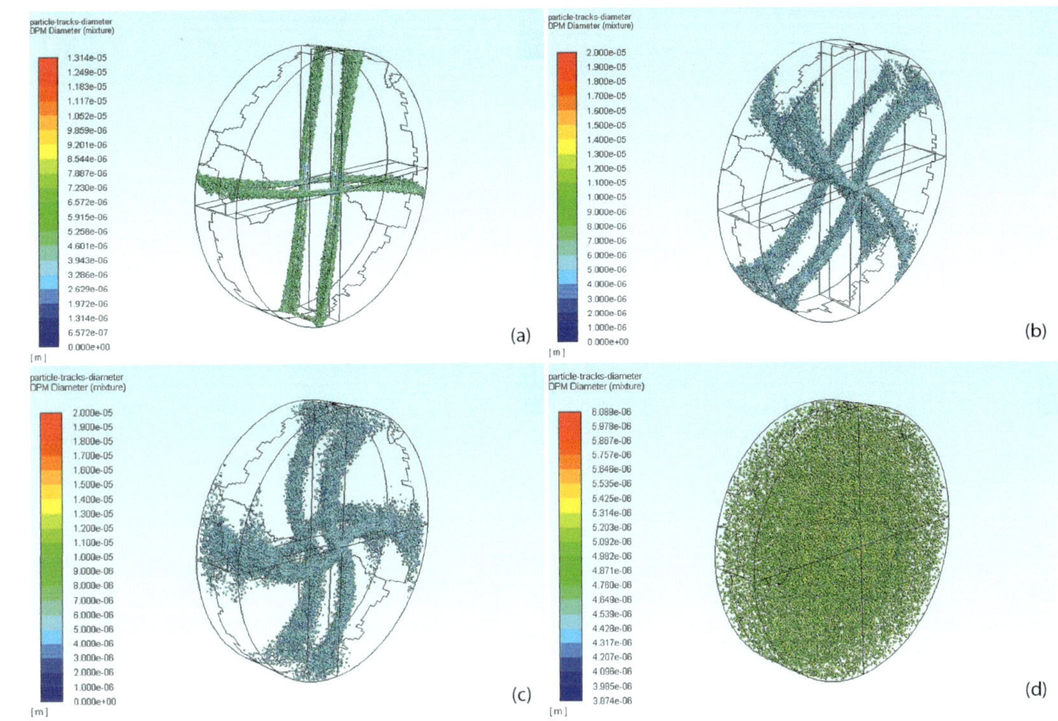

FIGURE 6.6 Spatiotemporal evolution of the instantaneous size distribution of ballotini particles $N = 70$ rpm, $e_{ss} = 0.99$, $d_{50,\text{ballotini}} = 10$ µm: (a) $t = 0.01$ s, (b) $t = 0.05$ s, (c) $t = 0.1$ s, and (d) $t = 1$ s.

6.4.4 Effect of Particle Restitution Coefficient on the Granular Pressure

In order to evaluate the effect of the particle normal restitution coefficients on the solids pressure, numerical experiments were performed at different particle restitution coefficients starting with a converged solution for the highest restitution coefficient. Figure 6.12 shows the profiles of the granular pressure for different restitution coefficients. The radial distribution of the solids pressure across the reactor diameter shows a region of high granular pressure near the rotating shaft with no significant variation between the restitution coefficients, with the exception of the lowest value of the restitution coefficient ($e = 0.2$). There is a sharp reduction in the values of the granular pressure with respect to the radial distance were observed close to the shaft of the reactor while away from this region, there is little of no differences in the values of the granular pressure for all the restitution coefficients. This observation is consistent with the results obtained from the numerical analysis of particle dispersion in stirred tank using two-fluid model [11]. Since the granular pressure at low solids concentration relies primarily on the kinetic mechanism rather than particle–particle collisions [15], this implies that discrete phase velocity does not seem to vary significantly with the restitution coefficients. The observed trend in the solids pressure with respect to the restitution coefficients might be attributed to the dilute nature of the flow under consideration. However, it is worth pointing out that a wider range of restitution coefficients are possible in practice due to particle roughness, and lower values of restitution coefficient are often encountered in practical applications. Gant and Gatzke [21] have reported restitution coefficients in the range of 0 to 0.33.

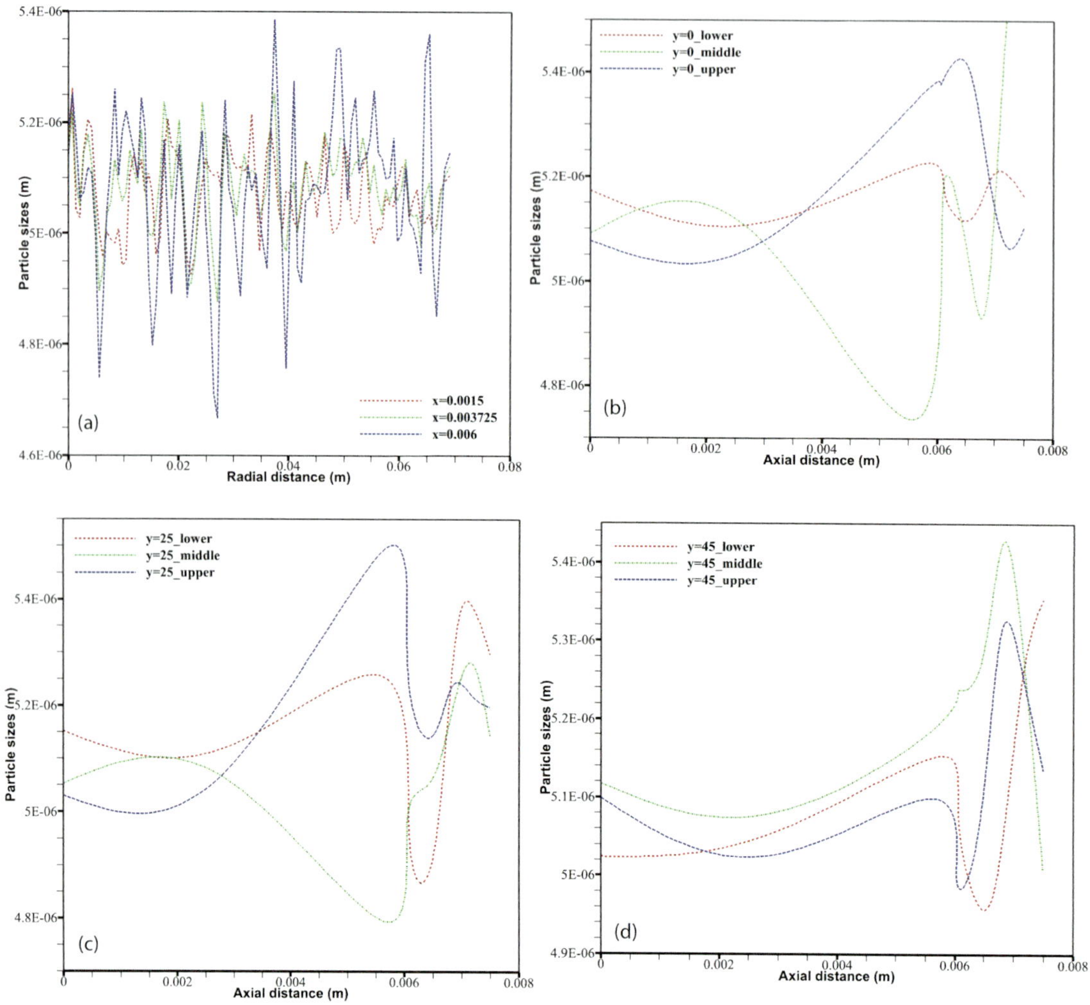

FIGURE 6.7 Vertical and horizontal profiles of the instantaneous size distribution of ballotini particles on the *YZ* and *XZ* planes after 1 s $N = 70$ rpm, $e_{ss} = 0.99$, $d_{50,\text{ballotini}} = 10$ μm: (a) vertical profiles, (b) horizontal profile $y = 0$, (c) horizontal profile $y = 0.025$, and (d) horizontal profile $y = 0.045$.

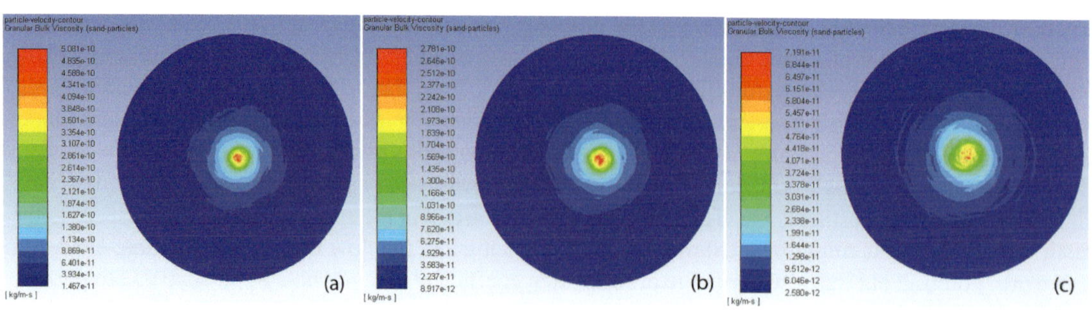

FIGURE 6.8 Contours of granular bulk viscosities of sand particles on the *YZ* plane after 1 s for different particle sizes $x = 0.0015$, $e_{ss} = 0.99$, $N = 70$ rpm: (a) 10 μm, (b) 6 μm, and (c) 2 μm.

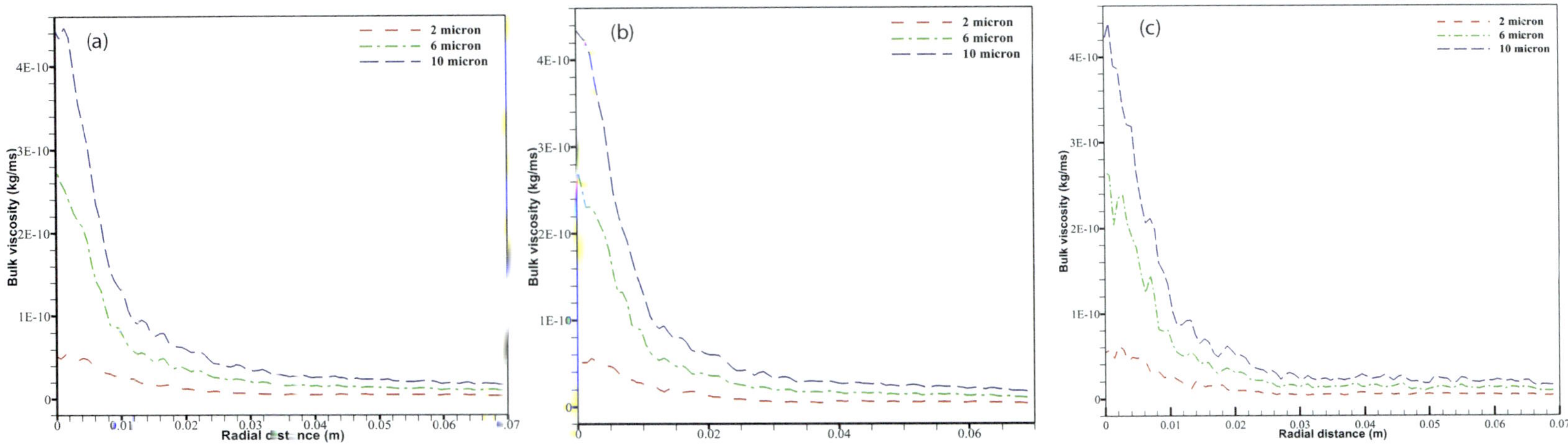

FIGURE 6.9 Bulk viscosity of sand particles as a function of radial distance for different particle sizes on the *YZ* planes after 1 s $e_{ss} = 0.99$: (a) $x = 0.0015$, (b) $x = 0.003725$, and (c) $x = 0.006$.

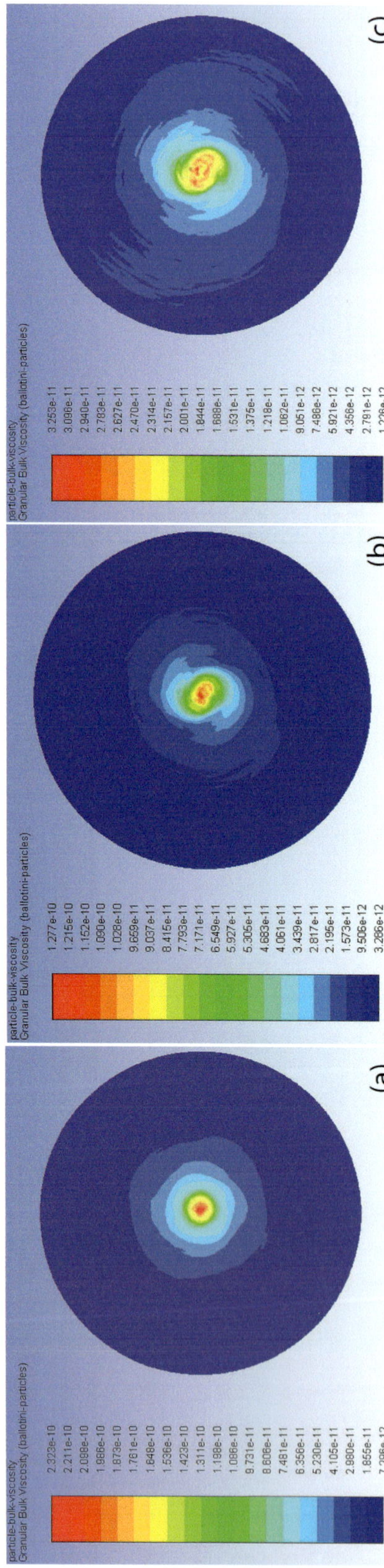

FIGURE 6.10 Contours of granular bulk viscosities of ballotini particles on the *YZ* plane after 1 s for different particle sizes $x = 0.0015$, $e_{ss} = 0.99$, $N = 70$ rpm: (a) 10μm, (b) 6 μm, and (c) 2 μm.

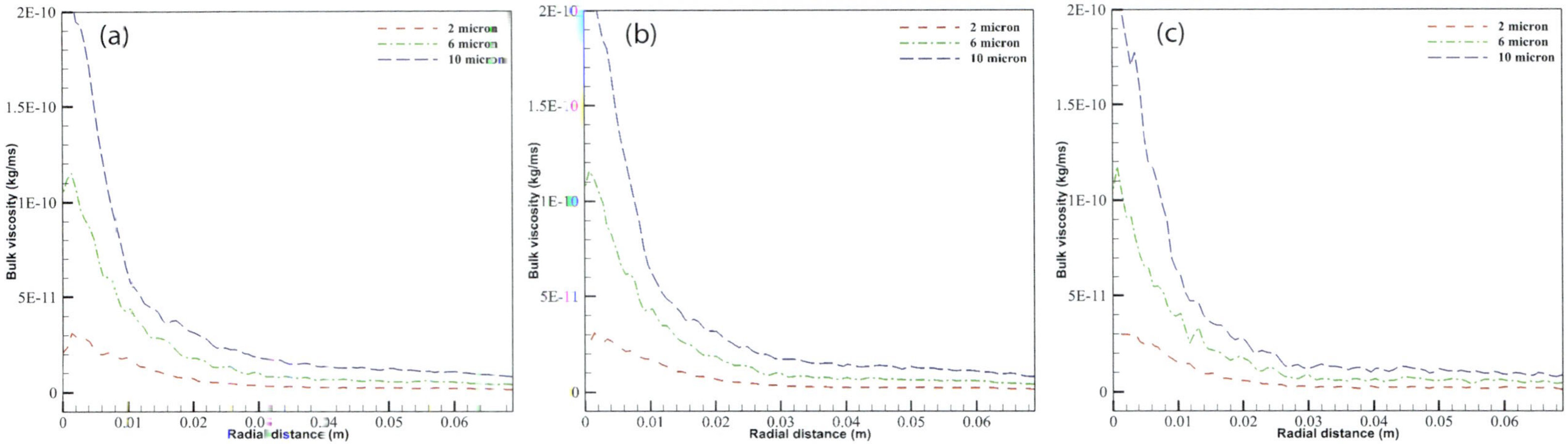

FIGURE 6.11 Bulk viscosity of ballotini particles as a function of radial distance for different mean particle sizes on the *YZ* planes after 1 s e_{ss} = 0.99: (a) x = 0.0015, (b) x = 0.003725, and (c) x = 0.006.

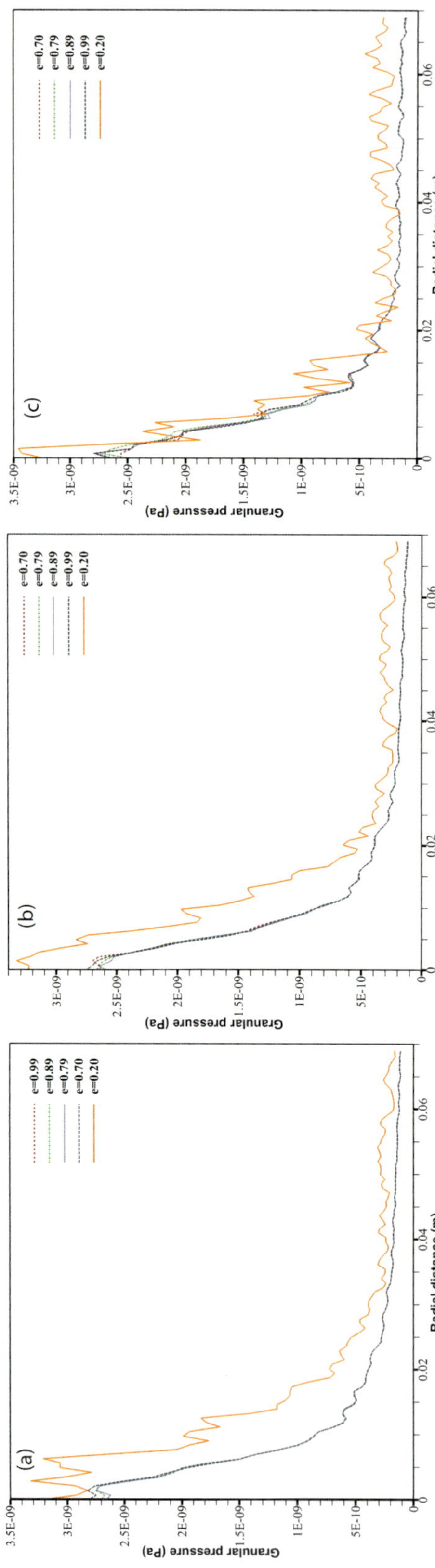

FIGURE 6.12 Granular pressure of sand particles as a function of radial distance for different particle restitution coefficients on the YZ planes after 1 s N = 70 rpm, $d_{p,sand}$ = 5 μm: (a) x = 0.0015, (b) x = 0.003725, and (c) x = 0.006.

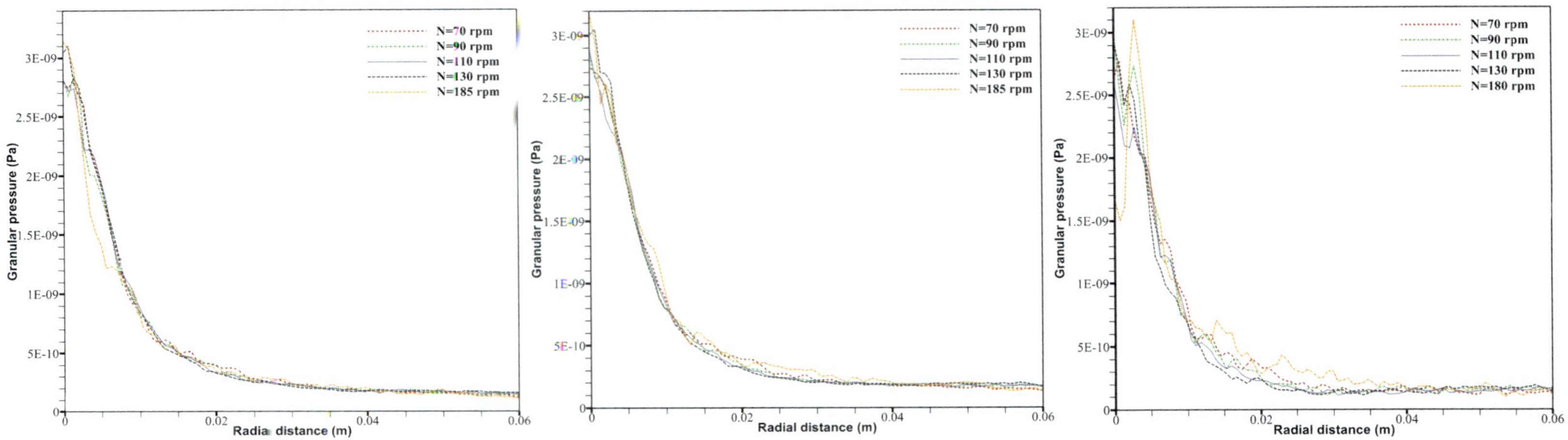

FIGURE 6.13 Granular pressure of sand particles as a function of radial distance for different rotation speeds on the *YZ* planes after 1 s $e_{ss} = 0.99$, $d_{p,sand} = 10$ μm: (a) $x = 0.0015$, (b) $x = 0.003725$, and (c) $x = 0.006$.

6.4.5 Effect of the Reactor Rotation Speed on the Granular Pressure

In assessing the effect of the reactor rotational speed on the granular properties, numerical simulations were performed at different operating speeds starting with the converged solution of the lowest operating speed of 70 rpm and gradually increasing the speed to the desired values. This is done to ensure stability of the numerical solution [24]. Figure 6.13 shows the influence of the rotational speed on the spatial distribution of the granular pressure. In terms of the temporal evolution of the solids pressure, a gradual reduction in its absolute values was observed for the *YZ* plane near the rotor as the particle tracking time increases and with the simulation approaching steady-state condition. The spatial distribution shows a consistent trend of high solids pressure around the shaft. This high solids pressure in this region might be due to the higher solids concentration in the area around the shaft resulting in a higher collisional contributions to the solids pressure [15]. The granular pressure in terms of the absolute values seem to correlate with higher rotational speed near the rotating shaft. However, away from this region, the solids pressure does not seem to vary significantly with the operating speed. One possible reason for this observation might be the reduced collisional effects due to lower solids concentration away from the shaft. One way to improve the homogeneity of the solids concentration is to incorporate baffles into this region to increase the vorticity, turbulence intensity, and consequently, particle dispersion.

6.5 CONCLUDING REMARKS

A coupled DPM–KTGF model was developed to study the interactions of granular materials within an enclosed Batchelor flow. The CFD model was employed to predict the spatiotemporal properties of the discrete particles as well as the flow characteristics using an in-house vortex reactor designed for the agglomeration of particles in suspension. Based on the findings of this study, the following conclusions were made:

- There is a spatiotemporal variation in the discrete phase concentration and particle size distribution (PSD) within the reactor, with higher particle concentration in the region around the shaft and a fairly uniform spatial distribution of the particle sizes.
- The granular bulk viscosity in terms of the absolute values exhibit a positive correlation with the particle sizes with increasing particle sizes resulting in higher bulk viscosities.
- The particle dispersity does not have any significant effect on the granular bulk viscosities which implies that the granular bulk viscosity is a function of the particle diameters rather than the dispersity of the particles.
- The restitution coefficients and the rotational speeds do not significantly influence the granular properties beyond the area close to the rotating shaft due to low turbulence intensity causing high solids hold up in this region.
- The granular pressure in terms of the absolute values have a positive correlation with higher rotational speed especially near the rotating shaft. However, away from this region, the solids pressure does not vary significantly with the operating speed.
- The granular pressure at low solids concentration relies primarily on the kinetic mechanism rather than particle–particle collisions which implies that discrete phase velocity does not vary significantly with the restitution coefficients, especially at low solids loading.
- The vertical velocity and vorticity profiles within the reactor give a reasonably good agreement between the CFPD model predictions and PIV measurements. The minor difference between CFPD model predictions and PIV measurements might be because some flow parameters were actually under-predicted by the PIV analysis.

Furthermore, the current CFPD model might also be extended to include other non-contact interparticle forces. Therefore, future work will focus on the extension of this model to other non-contact forces as well as other fluid–particle forces which were not considered in the present investigation using the square-well approach presented by Weber and Hrenya [22] or by full coupling of the flow field to the DEM model. It will also be interesting to study the full mechanisms of fluid–particle, particle–particle, and particle–wall interactions leading to particle collisions, coalescence/aggregation, growth, and breakage and compare the results of this fully resolved approach to the present investigation. The full DEM simulations, for instance, can be used to develop constitutive models to improve the KTGF model predictions. Ultimately, this approach will serve as a general framework for the modelling of complex fluid–particle interactions in high shear reactors and mixing vessels with rotating flows.

DATA AVAILABILITY

The data that support the findings of this chapter are available from the corresponding author, [B.O.], upon reasonable request.

NOMENCLATURE

Latin letters

Variable	**Description, Units**
p	Static pressure, Pa
p_s	Solids pressure, Pa
$\vec{v}$	Fluid velocity, m/s
u'	Fluctuating velocity component, m/s
$\overline{v_s}$	Particle velocity, m/s
$\vec{g}$	Gravitational force, m/s^2
$\vec{S}_{m,d_p}$	Interphase momentum exchange, kg/(m^2 s^2)
C_D	Drag coefficient, —
C_{D0}	Drag coefficient without turbulence modification, —
Re	Reynolds number, —
t	Flow time, s
k	Mean turbulent kinetic energy per unit mass, m^2/s^2
Δt	Change in time, s
F_D	Drag force acting on a single particle, N
F_P	Pressure gradient force acting on a single particle, N
F_{VM}	Virtual mass force acting on particles per unit mixture volume, N/m^3
F_G	Body force due to gravity acting on a single particle, N
F_B	Buoyancy force acting on a single particle, N
F_L	Lift force acting on particles per unit mixture volume, N/m^3
F_{int}	Particle–particle interaction force, N
F_{int}	Particle–particle interaction force, N
I_p	Moment of inertia, kg m^2

d_p/d_s	Particle or solids diameter, m
C_ω	Rotational drag coefficient, —
$\vec{T}$	Torque applied to particle by the fluid motion, Nm
e_{ss}	Coefficient of restitution for particle–particle collision, —
$k_{\Theta s}$	Turbulent mass diffusion coefficient, m^2/s

Greek Symbols

Variable	**Description, Units**
ρ	Density of the mixture, kg/m^3
ρ_f	Fluid density, kg/m^3
ρ_p	Particle density, kg/m^3
ρ_s	Particle or solids density, kg/m^3
μ	Dynamic viscosity, kg/m s
μ_{eff}	Effective viscosity, kg/m s
μ_t	Turbulent or eddy viscosity, kg/m s
$\{\mu_m\}$	Molecular viscosity, kg/(m s)
ε	Mean dissipation rate of kinetic energy, m^2/m^3
C_μ	Empirical model constant, —
$\vec{\omega}_p$	Particle angular velocity or rate of rotation, s^{-1}
$\vec{\Omega}$	Relative fluid–particle angular velocity, s^{-1}
$\vec{\tau}_s$	Solids stress tensor, N/m^2
α_s	Solids volume fraction, —
$\alpha_{s,\max}$	Solids volume fraction, —
μ_s	Shear viscosity, kg/(m s)
$\mu_{s,col}$	Collisional shear viscosity, kg/(m s)
$\mu_{s,kin}$	Kinetic shear viscosity, kg/(m s)
$\mu_{s,fr}$	Frictional shear viscosity, kg/(m s)
λ_s	Bulk viscosity, kg/(m s)
Θ_s	Solids granular temperature, m^2/s^2
$\gamma_{\Theta s}$	Energy dissipation rate in terms of the granular temperature, kg/(m s^3)
φ_{ls}	Dissipation of granular energy due to phase exchange, kg/m s^3
σ	Root mean square value of the fluctuating velocity, m/s
σ_ε	Empirical dispersed turbulence model constant, —
σ_k	Empirical dispersed turbulence model constant, —

Acronyms

DPM	Discrete phase model
DDPM	Dense discrete phase model
KTGF	Kinetic theory of granular flow
DEM	Discrete element model
PBM	Population Balance Model

CFPD Computational fluid–particle dynamics
DLVO Derjaguin–Landau–Verwey–Overbeek
RANS Reynolds-averaged Navier–Stokes
PIV Particle image velocimetry

REFERENCES

1. F. Chaumeil, M. Crapper, Using the DEM-CFD method to predict Brownian particle deposition in a constricted tube, *Particuology* 15 (2014) 94–106. https://doi.org/10.1016/j.partic.2013.05.005.
2. A. Haghnegahdar, Y. Feng, X. Chen, J. Lin, Computational analysis of deposition and translocation of inhaled nicotine and acrolein in the human body with e-cigarette puffing topographies, *Aerosol Sci. Technol.* 52 (2018) 483–493. https://doi.org/10.1080/02786826.2018.1447644.
3. F. Yu, Z. Xu, A. Haghnegahdar, Computational fluid-particle dynamics modeling for unconventional inhaled aerosols in human respiratory systems, in: K. Volkov (Ed.), *Aerosols*, InTech, Rijeka, 2016. https://doi.org/10.5772/65361.
4. F. Greifzu, C. Kratzsch, T. Forgber, F. Lindner, R. Schwarze, Assessment of particle-tracking models for dispersed particle-laden flows implemented in OpenFOAM and ANSYS FLUENT, *Eng. Appl. Comput. Fluid Mech.* 10 (2016) 30–43. https://doi.org/10.1080/19942060.2015.1104266.
5. D.O. Njobuenwu, M. Fairweather, Large eddy simulation of particle agglomeration with shear breakup in turbulent channel flow, *Phys. Fluids* 30 (2018). https://doi.org/10.1063/1.5037174.
6. J.J. Derksen, Numerical simulation of solids suspension in a stirred tank, *AIChE J.* 49 (2003) 2700–2714. https://doi.org/10.1002/aic.690491104.
7. J. Tu, G.-H. Yeoh, C. Liu, Y. Tao, *Computational Fluid Dynamics: A Practical Approach*, 4th ed., Elsevier, Oxford, 2023. https://doi.org/10.1016/C2021-0-01771-5.
8. G. Bellani, E.A. Variano, Slip velocity of large neutrally buoyant particles in turbulent flows, *New J. Phys.* 14 (2012). http://dx.doi.org/10.1088/1367-2630/14/12/125009.
9. D. Schellander, *CFD Simulations of Particle Laden Flow: Particle Transport and Separation*, Anchor Academic Publishing, Hamburg, 2014.
10. H.R. Norouzi, R. Zarghami, R. Sotudeh-Gharebagh, N. Mostoufi, *Coupled CFD-DEM Modeling: Formulation, Implementation and Application to Multiphase Flows*, John Wiley & Sons, Chichester, 2016. https://doi.org/10.1002/9781119005315.
11. L. Xie, Z.-H. Luo, Modeling and simulation of the influences of particle-particle interactions on dense solid–liquid suspensions in stirred vessels, *Chem. Eng. Sci.* 176 (2018) 439–453. https://doi.org/10.1016/j.ces.2017.11.017.
12. A. Darelius, A. Rasmuson, B. van Wachem, I. Niklasson Björn, S. Folestad, CFD simulation of the high shear mixing process using kinetic theory of granular flow and frictional stress models, *Chem. Eng. Sci.* 63 (2008) 2188–2197. https://doi.org/10.1016/j.ces.2008.01.018.
13. A. Hashemisohi, L. Wang, A. Shahbazi, Dense discrete phase model coupled with kinetic theory of granular flow to improve predictions of bubbling fluidized bed hydrodynamics, *KONA Powder Part. J.* 36 (2019) 215–223. https://doi.org/10.14356/kona.2019017.
14. J.K. Furtney, F. Zhang, Y. Han, Review of methods and applications for incorporating fluid flow in the discrete element method, in: *The Fifth International FLAC/DEM Symposium*, Itasca International Inc., Hangzhou, China, 2013.
15. S. Cloete, S.T. Johansen, S. Amini, Performance evaluation of a complete Lagrangian KTGF approach for dilute granular flow modelling, *Powder Technol.* 226 (2012) 43–52. https://doi.org/10.1016/j.powtec.2012.01.010.
16. E. Michaelides, C.T. Crowe, J.D. Schwarzkopf, *Multiphase Flow Handbook*, 2nd ed., CRC Press, Boca Raton, FL, 2016. https://doi.org/10.1201/9781315371924.
17. *ANSYS Fluent Theory Guide 2024R1* (2024).
18. D. Wadnerka, R.P. Utika, M.O. Tade, V.K. Pareek, CFD simulation of solid–liquid stirred tanks, *Adv. Powder Technol.* 23 (2012) 445–453. https://doi.org/10.1016/j.apt.2012.03.007.
19. S. Balachandar, J.K. Eaton, Turbulent dispersed multiphase flow, *Annu. Rev. Fluid Mech.* 42 (2009) 111–133. https://doi.org/10.1146/annurev.fluid.010908.165243.

20. B. Oyegbile, M. Hoff, M. Adonadaga, B. Oyegbile, Experimental analysis of the hydrodynamics, flow pattern and wet agglomeration in rotor-stator vortex separators, *J. Environ. Chem. Eng.* 5 (2017) 2115–2127. https://doi.org/10.1016/j.jece.2017.04.016.
21. J.A. Gannt, E.P. Gatzke, Kinetic theory of granular flow limitations for modeling high-shear mixing, *Ind. Eng. Chem. Res.* 45 (2006) 6721–6727. https://doi.org/10.1021/ie051267f.
22. M.W. Weber, C.M. Hrenya, Square-well model for cohesion in fluidized beds, *Chem. Eng. Sci.* 61 (2006) 4511–4527. https://doi.org/10.1016/j.ces.2006.02.008.
23. B. Oyegbile, G. Akdogan, Hydrodynamic characterization of physicochemical processes in stirred tanks and agglomeration reactors, in: O.M. Basha, B.I. Morsi (Eds.), *Chemical Engineering Laboratory Unit Operations and Experimental Methods*, InTech, London, 2018: pp. 57–77. https://doi.org/10.5772/intechopen.77014.
24. *ANSYS Fluent User's Guide 2024R1* (2024).
25. A. Taghizadeh, S.H. Hashemabadi, E. Yazdani, S. Akbari, Numerical analysis of restitution coefficient, rotational speed and particle size effects on the hydrodynamics of particles in a rotating drum, *Granul. Matter* 20 (2018) 1–13. https://doi.org/10.1007/s10035-018-0813-2.
26. B. Oyegbile, G. Akdogan, M. Karimi, Experimental and CFD studies of the hydrodynamics in wet agglomeration process, *ChemEngineering* 2 (2018) 32–49. https://doi.org/10.3390/chemengineering2030032.
27. R.N.P. Childs, *Rotating Flow*, Elsevier, Oxford, 2011. https://doi.org/10.1016/C2009-0-30534-6.
28. T. Bodnár, G.P. Galdi, Š. Nečasová, eds., *Particles in Flows*, Springer, Heidelberg, 2017. https://doi.org/10.1007/978-3-319-60282-0.
29. Z. Li, J.J. Derksen, Z. Gao, Models and applications for simulating turbulent solid–liquid suspensions in stirred tanks, *J. Chem. Eng. Japan* 48 (2015) 329–336. https://doi.org/10.1252/jcej.14we056.
30. A. Hobbs, Design and optimization of a vortex particle separator for a hot mix asphalt plant, in: *International ANSYS Conference*, ANSYS Inc., Pittsburgh, PA, 2004.
31. D. Wadnerkar, M.O. Tade, V.K. Pareek, R.P. Utikar, CFD simulation of solid–liquid stirred tanks for low to dense solid loading systems, *Particuology* 29 (2016) 16–33. https://doi.org/10.1016/j.partic.2016.01.012.
32. T. Bodnár, G.P. Galdi, Š. Nečasová, eds., Particles in Flows, Springer, Heidelberg (2017). https://doi.org/10.1007/978-3-319-60282-0.

Numerical Simulation of Convective Mixing and Heat Transfer in a Pilot-Scale Hydrothermal Carbonisation (HTC) Reactor

7

Benjamin Oyegbile

7.1 INTRODUCTION TO HYDROTHERMAL CARBONISATION PROCESS

Waste recycling and circular economy are two particularly important concepts that are fast gaining popularity in the environmental engineering field. This stems from the global waste management crisis and the associated negative environmental impact, thus necessitating the need for a more sustainable approach to waste management. Modern wastewater treatment is essential for public health and environmental protection, but it generates an unavoidable and complex by-product in the form of sewage sludge. Sewage sludge is a challenging waste stream due to its high moisture content and complex composition, which includes a high content of organic carbon, essential nutrients, and undesirable pollutants such as heavy metals and persistent organic compounds [1, 2]. The volume and nature of this waste pose a significant and costly techno-environmental challenge, with disposal costs estimated to account for 50–60% of the total operating costs of a wastewater treatment plant. The limitations of conventional sludge management practices underscore the urgent need for innovative solutions. Landfilling is becoming increasingly impractical due to land scarcity and stringent environmental regulations. Incineration, while effective at reducing sludge volume, requires substantial energy for pre-drying to achieve thermally efficient conditions, which is

DOI: 10.1201/9781003518426-7

the point at which the process can sustain itself without external heat input. This reliance on pre-drying makes it an energy-intensive process and results in the loss of valuable carbon and nutrients. Anaerobic digestion (AD), a widely used biological process, is effective at producing biogas from organic matter but can only biodegrade approximately half of the organic content in the sludge, leaving a considerable amount of difficult-to-manage digestate that still requires further treatment. In response to these challenges, significant scientific efforts have been expended in addressing this problem with mixed results. Hydrothermal carbonisation has emerged as a promising alternative. Hydrothermal carbonisation (HTC) of wet biomass offers huge potential in the quest for sustainable waste management, given that the process can yield valuable solid and liquid products. The main advantage of the HTC process lies in its ability to convert wet biomass into useful products without prohibitive energy demand that is often associated with other forms of thermal energy conversion technologies, such as pyrolysis and incineration. HTC is a thermochemical process that operates on wet biomass, a distinct advantage over conventional thermal conversion methods. It is designed to overcome the shortcomings of traditional processes by efficiently converting waste biomass such as sludge into a more valuable material while simultaneously addressing issues of dewaterability, volume reduction, and pathogen destruction [3].

Hydrothermal carbonisation is a wet thermochemical conversion process that leverages the unique properties of subcritical water. It is an exothermic process and the resulting hydrochar is a dense solid with very high heating value, making it a viable replacement for coal as a cheap source of energy [4]. The process involves heating a moist biomass feedstock, such as sewage sludge and agricultural waste with a typical moisture content of 10–60%, in a closed pressurised vessel. The operating conditions are typically set at temperatures ranging from 180 to 260°C and corresponding autogenous pressures of 1.5–3.0 MPa (approximately 10–40 bar). These conditions are crucial, as they keep the water in its liquid, subcritical state, preventing it from boiling. The feedstock residence time within the reactor can vary from 10 min to several hours, depending on the desired product characteristics. During this process, the biomass is transformed into three primary products: a solid, carbon-rich hydrochar; an aqueous slurry referred to as process water; and a small quantity of gas, primarily composed of CO_2 (from decarboxylation) with minor H_2, CH_4, and contaminants (H_2S, NO_x, NH_3), which are typically removed by scrubbing. Heat is supplied by steam or thermal oil, and the process can be exothermic once initiated with the continuous recirculation of steam and the hydrochar, ensuring an autothermal operation. The process is typically operated in batch mode, but there is a gradual shift towards semi-batch and continuous systems to improve the overall yield of the hydrochar. When the HTC reactor is operated at an optimum condition, a hydrochar yield of 20–60% of dry mass has been reported from sewage sludge. The resulting hydrochar has heating values of approximately 20–25 MJ/kg DS [5, 6].

The conversion of wet biomass during HTC is a result of a series of complex, interdependent chemical reactions that imitate the natural coal formation process over a vastly compressed timeframe. These reactions occur under high temperature and pressure and are responsible for the transformation of wet biomass into a more stable, carbon-rich material. A central and defining aspect of the HTC process is the use of water as the reaction medium. In this environment, water acts as a solvent, a reactant, and an efficient heat transfer medium. The fundamental advantage of this approach is that it circumvents the need for energy-intensive pre-drying of the feedstock, a step that is of significant energy and cost burden for other thermochemical conversion processes like pyrolysis and gasification. By capitalising on the water that is already inherent in the sludge, HTC offers a more energy-efficient pathway for waste valorisation with the following reactions occurring during the transformation of sludge into biochar [5]:

- Hydrolysis: This is the initial stage where complex organic biopolymers in the sludge, such as proteins, polysaccharides, and lipids, are degraded into simpler, water-soluble monomers and oligomers.
- Dehydration and decarboxylation: These reactions are pivotal for the densification of carbon. Dehydration involves the removal of water molecules, while decarboxylation liberates carboxyl groups, resulting in the formation of CO_2. These reactions play a vital role in lowering the hydrogen-to-carbon (H/C) and oxygen-to-carbon (O/C) atomic ratios, which is characteristic of the production of a carbon-rich biochar.

- Polymerisation and aromatisation: Following hydrolysis, the liberated monomers and oligomers undergo recombination and condensation reactions. This process leads to the formation of a more stable, solid biochar with enhanced aromaticity and hydrophobic.

The final properties and yield of the final products are highly dependent on the chosen process parameters. Temperature and residence time are the most influential parameters, and there is a fundamental trade-off depending on the desired end product quality. Higher temperatures and longer residence times lead to a lower hydrochar yield but produce a material with higher carbon content and higher heating value (HHV). This is due to the increased severity of the process, which promotes the decomposition of organic matter, with a greater proportion breaking down and volatilising into the aqueous and gas phases. Conversely, less severe conditions (lower temperature and shorter residence time) result in a higher biochar yield but with lower carbon content and HHV. This causal relationship presents a critical decision point for HTC operators when deciding whether the process should be optimised for maximising the volume of hydrochar for applications like soil amendment, or for maximising energy content when the target application is for use as an energy source. The addition of chemical additives, such as sulfuric acid or alkali, can also influence the process chemistry. For instance, acidic conditions can increase the hydrochar's carbon content and thermal stability. Furthermore, the addition of acids can promote the migration of heavy metals and phosphorus into the liquid phase, a critical factor for subsequent nutrient recovery.

In terms of the development and scale-up of HTC reactors, a quick survey of the scientific archives reveals many scientific publications devoted to the study and development of HTC reactors for converting different types of feedstocks to valuable products [1, 2, 4, 6–8]. However, the majority of these studies do not provide any clear path to scale-up or industrial uptake and are mainly for pure academic purposes rather than any useful commercial venture. Mendecka et al. [9] conducted a CFD study on the thermofluid dynamics of the batch HTC processing of olive pomace using a mixture multiphase model in ANSYS Fluent. They employed a first-order reaction kinetic model based on the work of Kneževic [10] coupled with the equation of fluid flow to predict the dynamics of the hydrothermal conversion process. However, contrary to other studies on the HTC process, they did not consider the effect of evaporation and condensation on the HTC process based on their assumption that water remains in the liquid phase throughout the process. Despite this assumption, the simulation case with a power rating of 4.0 kW showed close agreement with the experimental data on the average temperature in the reactor and was used to validate the model. In a related study, Chater [11] and co-workers carried out a 2D numerical analysis of the fluid flow and heat transfer in a batch reactor for the processing of raw olive pomace (ROP). A basic kinetic model to predict the hydrothermal decomposition of the biomass coupled with fluid flow equation was implemented numerically using COMSOL Multiphysics CFD software. The predictions of the CFD model reported showed a close agreement with average temperature and pressure measurements from experimental analysis of the HTC process performed under similar operating conditions.

This chapter seeks to demonstrate the application of engineering simulation in the scale-up and development of HTC reactor by conducting an applied computational study on the application of hydrothermal carbonisation process to the treatment of biological waste such as sewage sludge. In performing this study, a two-phase convective heat and mass transfer simulation was performed to study the expected thermophysical conditions in the reactor. The numerical solution was performed locally and on High Performance Computing (HPC) facility using a commercial software package (ANSYS Fluent) extended with custom user-defined functions (UDFs) for the thermophysical properties. The study was part of a larger project on the designing, testing, and optimisation of a prototype pilot-scale HTC reactor that meets high technical sustainability and economic criteria, such as robust process performance, high product quality, and low operating cost. A comprehensive investigation into the fluid flow dynamics in the (HTC) reactor was conducted using the computational platform with a view to understand and assess the initial process performance. Furthermore, it is envisaged that a deep understanding of the process dynamics at this stage will help in identifying the different aspects of the process that need to be optimised. Using the CFD platform to implement this study provides a flexible platform to investigate different 'what-if' scenarios and will significantly complement any experimental work and scaling-up process.

7.2 METHODOLOGY FOR THE CFD MODEL DEVELOPMENT

A multiphase volume of fluid (VOF) CFD model was implemented for this investigation using ANSYS Fluent v2024R1 (ANSYS Inc.). VOF is a very popular multiphase modelling approach due to its accuracy in tracking the vapour–liquid interface and is ideally suited to a system of immiscible fluids such water–gas or water–oil mixtures [12]. The flow domain shown in Figure 7.1 represents the 2D model of a pilot-scale (HTC) reactor with a structured quadrilateral mesh, while the investigated operating conditions are given in Table 7.1. The HTC reactor, which is initially half-filled with water at the beginning of the numerical experiment and contains air at a temperature of 333 K and atmospheric pressure of 101.325 kPa, was heated through the side walls. The top and lower sections of the reactor are not insulated and therefore experience heat loss to the ambient environment by convective dissipation. Reynolds-averaged Navier–Stokes (RANS) simulations were performed with turbulence modelling using an algebraic volume of fluid (VOF) scheme to account for the multiphase aspect of the flow. Appropriate settings such as SIMPLEC for the pressure–velocity coupling, and second-order upwind spatial discretisation scheme for all other variables, with the exception of pressure for which PRESTO is used to capture the pressure build-up in the reactor as well as the associated phase change (evaporation/condensation) as the temperature

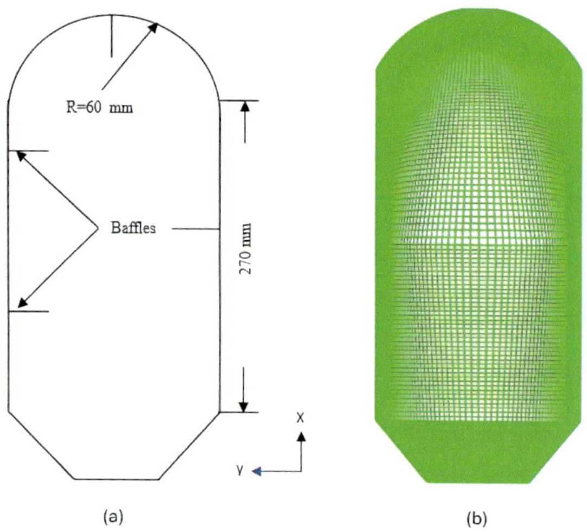

FIGURE 7.1 The reactor geometry and the flow domain: (a) CFD domain (not to scale) and (b) computational grid.

TABLE 7.1 Specification for the Temperature and Pressure-Dependent Fluid Material Properties

	FLUID TYPE	
PROPERTIES	*LIQUID (WATER)*	*GAS (AIR, VAPOUR)*
ρ (kg/m^3)	$95.908 + 9.063T - 3.217 \times 10^{-2}T^2 + 4.87 \times 10^{-5}T^3 - 2.96 \times 10^{-8}T^4$	Ideal gas law
C_p (J/kg K)	Linear interpolation of data points from the Fluent database	Mixing law from the Fluent database
k (W/m K)	$-0.695 + 7.844 \times 10^{-3}T - 1.360 \times 10^{-5}T^2 + 6.434 \times 10^{-9}T^3$	0.0454
μ (kg/m s)	$0.096 - 1.033 \times 10^{-3}T + 4.469 \times 10^{-6}T^2 - 9.660 \times 10^{-9}T^3 + 1.042 \times 10^{-11}T^4 - 4.478 \times 10^{-15}T^5$	1.72×10^{-5}
T_{sat} (K)	$318.649 + 2.923 \times 10^{-4}P - 1.799 \times 10^{-10}P^2 + 3.493 \times 10^{-17}P^3$	—
σ (—)	$0.082 + 1.141 \times 10^{-4}T - 5.985 \times 10^{-7}T^2 + 3.339 \times 10^{-10}T^3$	—

of the fluid increases. The second-order implicit temporal discretisation scheme was used for the time advancement along with temperature and pressure-dependent coefficients for the surface tension force and saturation temperature. In addition, temperature and pressure-dependent material properties were specified for all the fluids enclosed within the reactor using the built-in Fluent and NIST fluid thermophysical database, as shown in Table 7.1.

The governing equations for the VOF approach originally proposed by Hirt and Nichols [13] consist of a set of mass and momentum equations for an incompressible flow combined with a separate transport equation for tracking the volume fractions of the secondary phase in the system (Equations 7.1–7.3). In addition, the surface tension force in the momentum equation was calculated from the continuum surface force (CSF) model proposed by Brackbill et al. [14], while a saturation–condensation model (Lee model) was chosen to capture the mass transfer from the primary to the secondary phase and vice versa due to the expected phase change in the vessel (Equations 7.4 and 7.5). Pressure-dependent saturation temperature was determined from the Antoine relationship given in Equation (7.6). Given that the Rayleigh numbers for the fluids will rise significantly due to high temperature and intense pressure in the vessel, an appropriate turbulence model (RNG k – ε) with scalable wall functions was selected owing to its ability to accurately predict natural convective flows at low computational cost, with special consideration for the high thermal and velocity gradients expected at the wall. The resolution of the computational grid used in the numerical computations was significantly refined near the wall via a grid independence study so as to sufficiently capture the expected high thermal and velocity gradients. A 2D axisymmetric model of the reactor was considered with the assumption that the flow is symmetric about the axis, which significantly lowers the computational cost.

A systematic workflow depicted in Figure 7.2 was followed in performing the numerical calculations. First, an axisymmetric model of the physical reactor was created using a computer-aided design (CAD) software (ANSYS design modeler) from which a numerical grid was generated using built-in ANSYS Mesher. Thereafter, the numerical grid was loaded into the numerical simulation software (ANSYS Fluent) and appropriate initial boundary and operating conditions presented in Table 7.1 were selected to mimic the physical thermofluidic conditions in the HTC reactor. Additional models such as the VOF, RNG k – ε, Lee and CSF models necessary to mimic the physics involved in the actual HTC process were also switched on after which the numerical computations were performed [12, 15–17]:

$$\frac{\partial \bar{v}_i}{\partial x_i} = 0 \tag{7.1}$$

$$\rho \frac{\partial \bar{v}_i}{\partial t} + \rho \bar{v} \frac{\partial \bar{v}_i}{\partial x_j} = -\frac{\partial \bar{p}}{\partial x_i} + \frac{\partial}{\partial x_j}\left(\tau_{ij} - \rho \overline{v'_i v'_j}\right) + \rho \vec{g} + F_\sigma \tag{7.2}$$

$$\frac{1}{\rho_q}\left[\frac{\partial}{\partial t}\left(a_q \rho_q\right) + \nabla\left(a_q \rho_q \bar{v}\right)\right] = \frac{1}{\rho_q}\left(\dot{m}_{pq} - \dot{m}_{qp}\right) \tag{7.3}$$

$$F_\sigma = 2\sigma \frac{\rho k \nabla \alpha}{\left(\rho_l + \rho_v\right)} \tag{7.4}$$

$$\dot{m}_{lv} = \begin{cases} \gamma_l a_l \rho_l \dfrac{T - T_{sat}}{T_{sat}} & if\, T > T_{sat} \\[2ex] \gamma_v a_v \rho_v \dfrac{T - T_{sat}}{T_{sat}} & if\, T < T_{sat} \end{cases} \tag{7.5}$$

$$T_{sat} = \frac{A}{B - \log_{10} P} - C \tag{7.6}$$

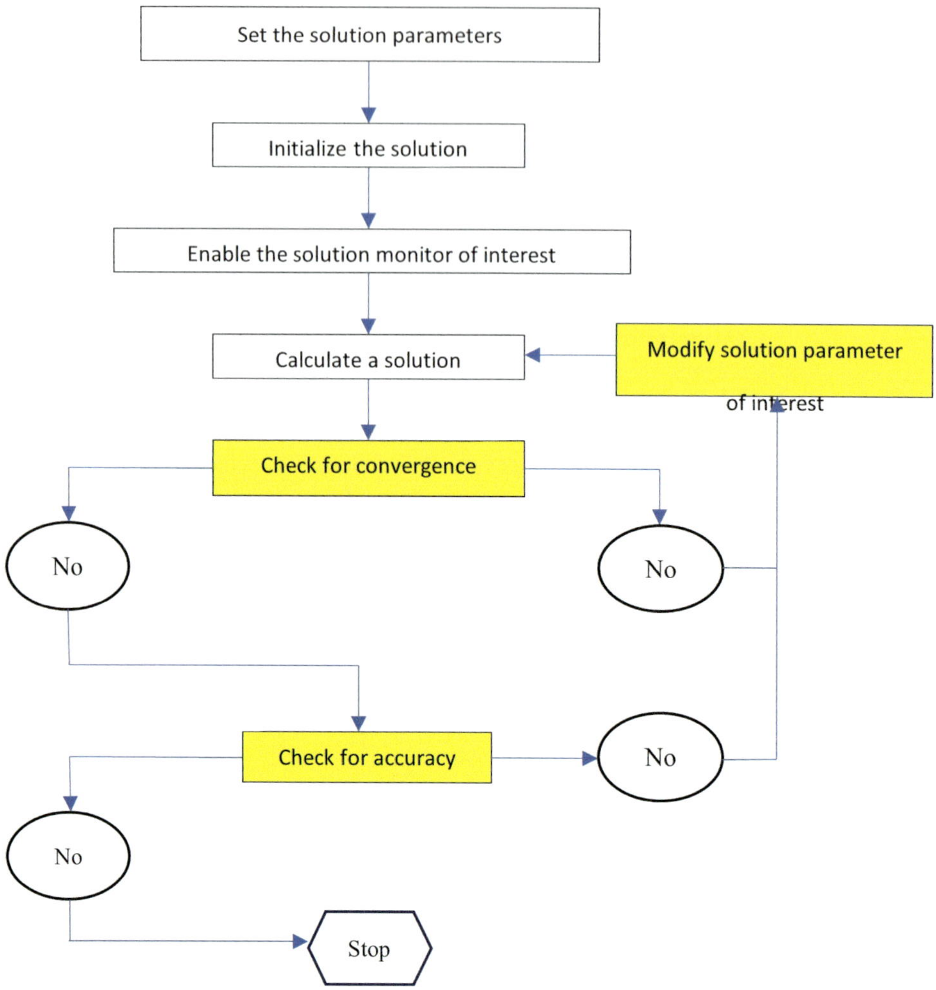

FIGURE 7.2 A flow chart illustrating the steps involved in the numerical simulations.

In Equations (7.1)–(7.6), T_{sat}, P is the partial water pressure, ρ is the density, $\bar{v}$ is the mean velocity, $\bar{p}$ is the mean pressure, $\bar{\tau}_{ij}$ is the viscous stress tensor, $\rho\overline{v_i'v_j'}$ is the Reynolds stresses which represent transfer of momentum due to turbulent fluctuations, t is the time, $\vec{g}$ is the gravitational body force acting in the negative x direction, F_σ is the surface tension force, σ is the surface tension coefficient, k is the surface curvature, γ_l and γ_v are the coefficients that determine the mass transfer rate, T_{sat} is the saturation temperature, while m_{lv}, m_{pq}, and m_{qp} represent the mass transfer from phase p (liquid) to phase q (vapour) and vice versa.

7.3 SOLUTION IMPLEMENTATION AND CONTROL

Two cases of the (HTC) reactor were investigated in batch mode with the operating conditions shown in Table 7.2. The fluid domain was modelled as an axisymmetric problem, and as such, only half of the fluid domain was modelled. A constant wall temperature boundary condition with a constant temperature of 250°C was imposed on the two lateral walls supplying heat into the vessel, while a convective heat

TABLE 7.2 Initial and Operating Conditions for the Two Test Cases Investigated in the Study

TEST CASE	*INITIAL FLUID T (K)*	*INITIAL WALL T (K)*	*INITIAL P (PA)*	*TIME (S)*
Test case 1	333	523	1,01,325	1800
Test case 2	333	573	1,01,325	9000

transfer boundary condition was imposed on the top and bottom wall with a heat transfer coefficient of 250 W/(m^2 K) and outside air temperature of 20°C. In the first test case, the simulation was performed for 30 min to study the initial reactor start-up, heat transfer, and evolution of the flow dynamics with the solution data analysed at shorter intervals. The second case involved a much longer simulation time of 120 min to evaluate the thermodynamics of the closed vessel beyond the initial start-up time and ascertain the point at which a steady-state condition is reached. The results of the two test cases along with the mesh sensitivity study, the solution accuracy, and validation checks are presented in the following sections.

7.3.1 Grid Sensitivity to Spatial Discretisation

An assessment of the optimum mesh resolution required for obtaining accurate results from the numerical simulations at a reasonable computational expense was carried out by conducting a mesh sensitivity study. This was done by a gradual increase of the mesh resolution and checking the corresponding effect on the predictions of the flow parameters. As the mesh density increases, a point is reached when an increase in the mesh resolution does not yield any significant change in the solution accuracy, and the mesh resolution at this point is deemed sufficient.

The results of the mesh sensitivity study are presented in Table 7.3 and Figure 7.3. It can be seen from Figures 7.3 that a successive increase in the mesh resolution from mesh 1 to mesh 4 has a significant effect on the predictions, but increasing the resolution further, as done in mesh 5, yields only a minor change in the predicted values of flow parameters. Therefore, a structured quad mesh with a resolution of 14,490 cells (mesh type 4) is deemed sufficient to obtain an acceptable result at a reasonable computational expense. This mesh resolution was subsequently used in computing the investigated fluid flow parameters. All computations were run locally using a Dell Latitude 7000 series laptop having a CPU with 8 cores and 16 GB

TABLE 7.3 Description of the Mesh Types and the Grid Resolution (Cell Count)

MESH 1	*MESH 2*	*MESH 3*	*MESH 4*	*MESH 5*
3240 cells	7260 cells	10,220 cells	14,490 cells	19,291 cells

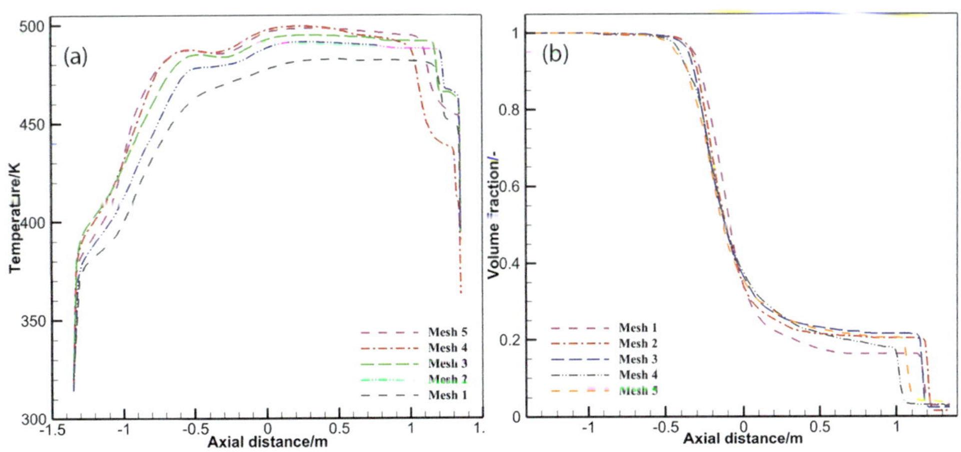

FIGURE 7.3 Axial flow profile of temperature and water volume fraction at $y/d = 0.5$ for different mesh resolutions after a total flow time of 1800 s: (a) temperature and (b) volume fraction.

memory and on an HPC cluster having a CPU with 32 cores and 64 GB memory for a total computation time of 24–120 h depending on the specified flow time and required level of solution accuracy.

7.3.2 Grid Sensitivities to Transient Discretisation

It is essential to ensure the robustness of the CFD predictions of the chosen mesh resolution, and this was achieved by performing a solution sensitivity test to compare the predictions of the selected mesh under different time-step sizes and transient temporal discretisation formulations. A higher discretisation scheme and shorter time-step size will undoubtedly produce more robust and accurate predictions, while the reverse is the case for a lower level of solution control. However, a high level of solution control is not justified in all cases, as the marginal increase in accuracy might be too prohibitive in terms of the computational cost. Therefore, it is necessary to perform a mesh sensitivity test to decide on the appropriate settings that meet the requirements of the investigation and the required level of accuracy. The results of the mesh sensitivity test to transient solution control methods are presented in Figures 7.4 and 7.5. The figures showed a marked improvement in the temperature predictions when the time-step size is increased from 0.05 s to 0.1 s. However, increasing the time-step size further yields only a marginal increase in temperature predictions with a disproportionate increase in computational cost. Therefore, time-step sizes of 0.05 s and 0.1 s for the first- and second-order transient formulations, respectively, were chosen as an acceptable compromise between CFD accuracy and computational demand.

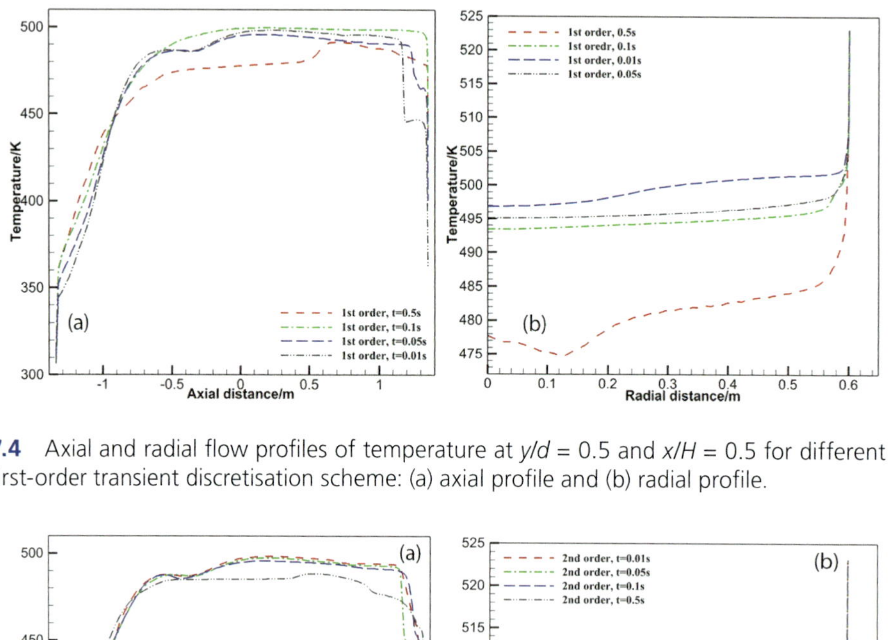

FIGURE 7.4 Axial and radial flow profiles of temperature at *y*/*d* = 0.5 and *x*/*H* = 0.5 for different time-step sizes for first-order transient discretisation scheme: (a) axial profile and (b) radial profile.

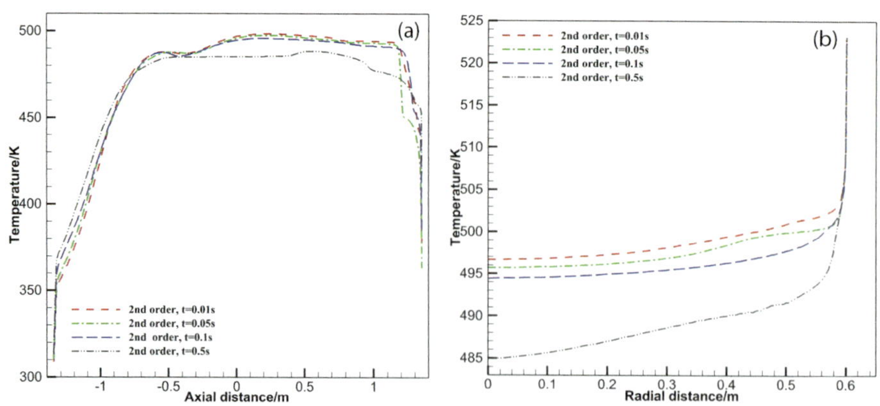

FIGURE 7.5 Axial and radial flow profiles of temperature at *y*/*d* = 0.5 and *x*/*H* = 0.5 for different time-step sizes for second-order transient discretisation schemes: (a) axial profile and (b) radial profile.

7.4 RESULTS AND DISCUSSION

7.4.1 Evolution of the Flow Dynamics and Heat Transfer

Figures 7.6–7.8 show the spatial and transient evolution of the flow parameters of interest for the initial transient phase of the simulation, while Figures 7.9–7.12 show the evolution of the fluid dynamics towards steady-state thermofluidic condition after approximately 120 min. It can be observed from the results of the first test case that there was a rapid increase in the fluid temperature under the prevailing isothermal conditions at the early stage of the simulation. The temperature in the bulk of the fluid, especially in the upper part of the reactor where there is a mixture of air and vapour, can be seen to be rising rapidly and approaching the wall temperature. This seems to suggest that there is a uniform temperature distribution in this zone, which is in sharp contrast to the lower section filled with water where there is thermal stratification. In the second test case in which the simulation was performed for a much longer flow time, it can be clearly seen that the flow parameters have reached a steady state with little or no change in the average flow condition with respect to the flow time. However, compared to the first test case, there was only a marginal increase in the temperature and water volume fraction values, which seems to suggest that most of the heat and mass transfer occurs in the first 30–60 min. A notable exception is the vapour–liquid interphase shown in Figure 7.10, which clearly shows a continued decrease in the water volume fraction due to evaporation well beyond the initial start-up time of 30 min. Table 7.4 shows the wall heat flux and heat transfer rate for the two test cases. Similar to the temperature fields, the wall heat flux and heat transfer rate indicate little to no change after 120 min, suggesting a steady-state condition has been reached.

In order to understand the changes in the thermodynamic state of the mixture in the closed vessel, important fluid properties such as the air, water, and vapour Rayleigh number, as well as the vapour-specific humidity, were monitored in the course of the simulation. Figure 7.13 shows the transient evolution of the monitored parameters for the second test case. Given that the thermodynamic state of the saturated mixture in the vapour zone is the primary driver of the HTC process, a good understanding of the temporal evolution of these parameters is particularly important in optimising the process performance. The Rayleigh number is an indication of the flow regime (laminar or turbulent), while the specific humidity which is a measure of the mass or volume of the water vapour compared to the total mass or volume of gas in the vessel, is an indication of the quality of the vapour with a value close to 1 indicating high vapour quality. Figure 7.13a shows that the flow in the vessel is fully turbulent after 30 min as all the fluid Rayleigh numbers are above the critical values, while the specific humidity shown in Figure 7.13b reaches a value close to 1.0 which indicates that the vapour is of high quality [18].

7.4.2 Solution Accuracy and Validation Check

The solution accuracy was checked against the theoretical calculations from the first principles by comparing the predicted net heat supplied to the vessel against the change in the internal energy of the fluids in the vessel. Values of the heating power were extracted directly from the Fluent CFD solver while the fluid thermophysical reference values were retrieved from the NIST database.

TABLE 7.4 Surface-Averaged Wall Heat Transfer Parameters

HEAT TRANSFER PARAMETERS	*FLOW TIME*			
	1800 (S)	*5400 (S)*	*7200 (S)*	*9000 (S)*
Wall heat flux (W/m^2)	17001.30	1640.57	15,598.23	15,109.44
Heat transfer rate (W)	135193.44	102278.72	10,0373.20	97,269.18

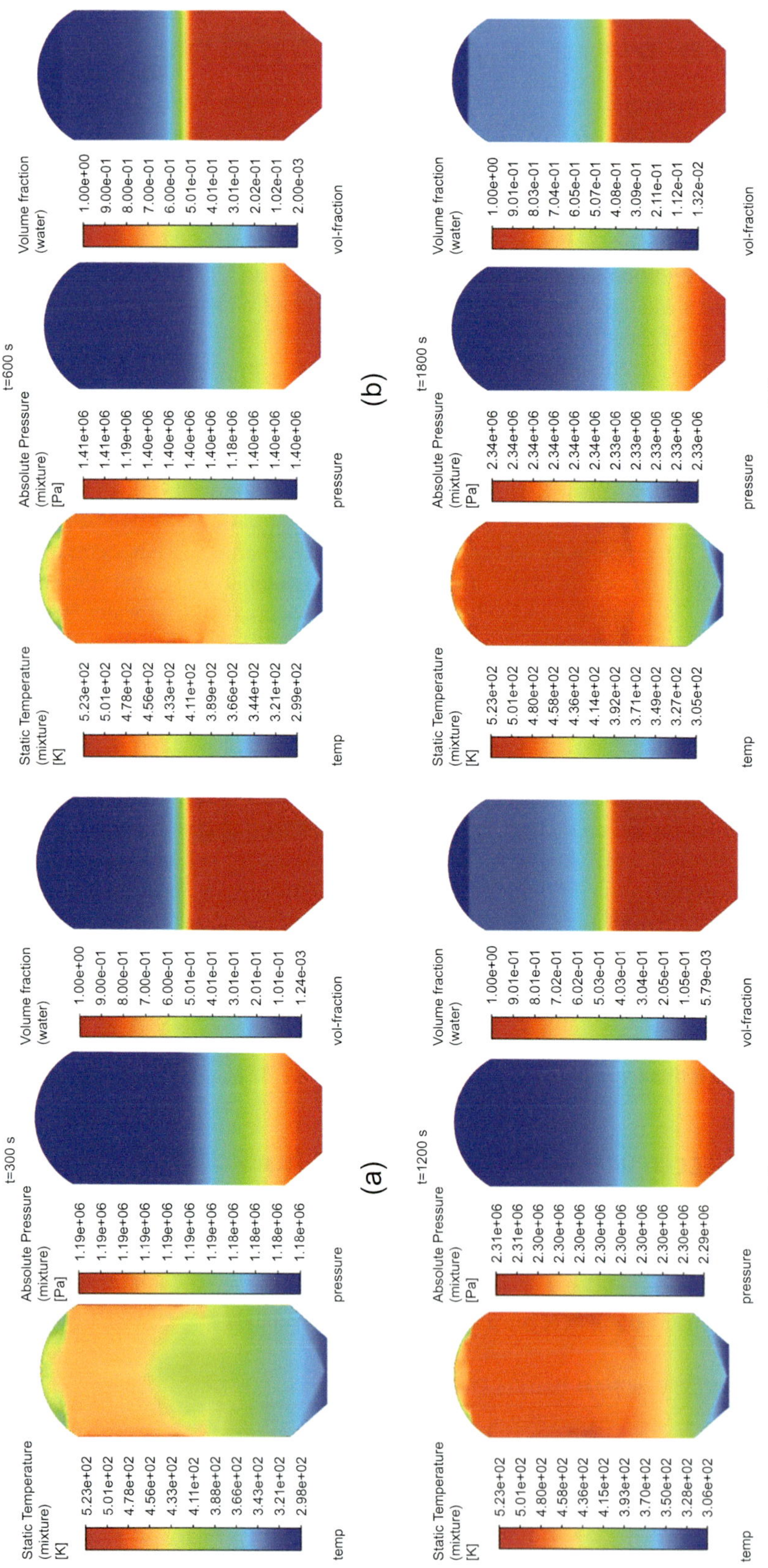

FIGURE 7.6 Contours of temperature, pressure, and water volume fraction at different flow times during the initial phase of the simulation: (a) $t = 300$ s, (b) $t = 600$ s, (c) $t = 1200$ s, and (d) $t = 1800$ s.

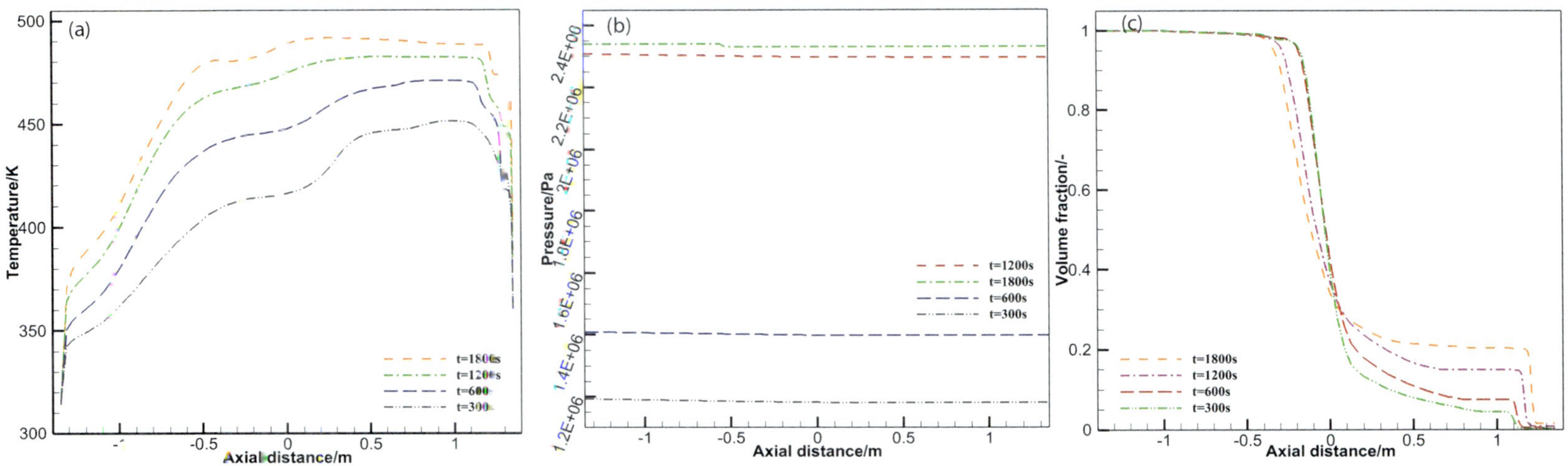

FIGURE 7.7 Axial flow profiles of temperature, pressure, and water volume fraction at $y/d = 0.5$ for different flow times: (a) temperature, (b) pressure, and (c) volume fraction.

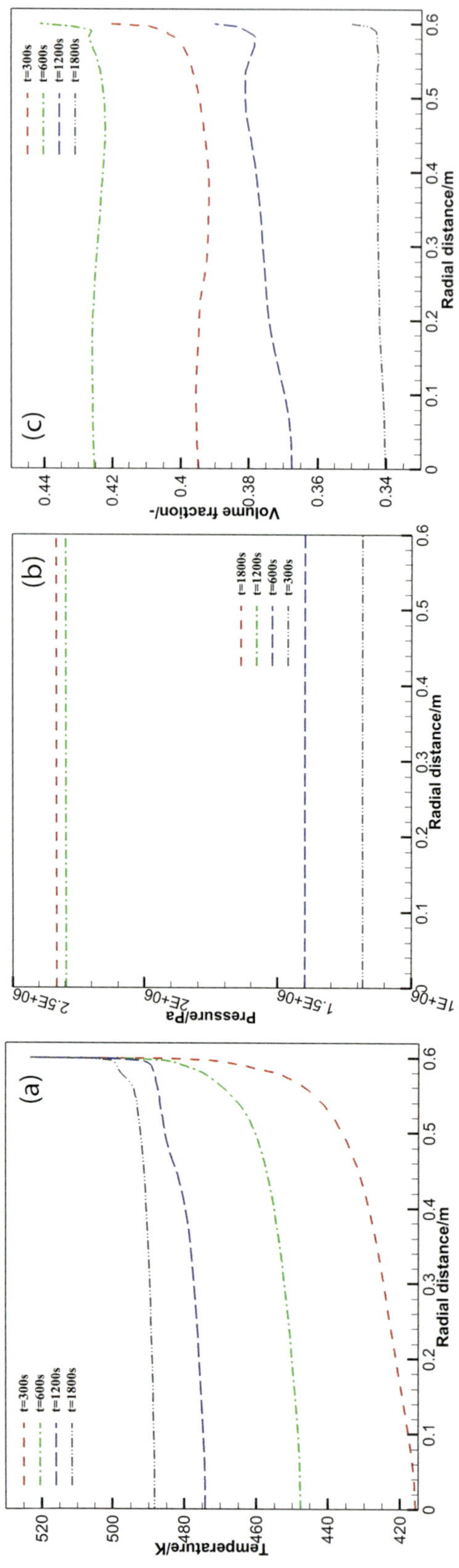

FIGURE 7.8 Radial flow profiles of temperature, pressure, and water volume fraction at $x/H = 0.5$ for different flow times: (a) temperature, (b) pressure, and (c) volume fraction.

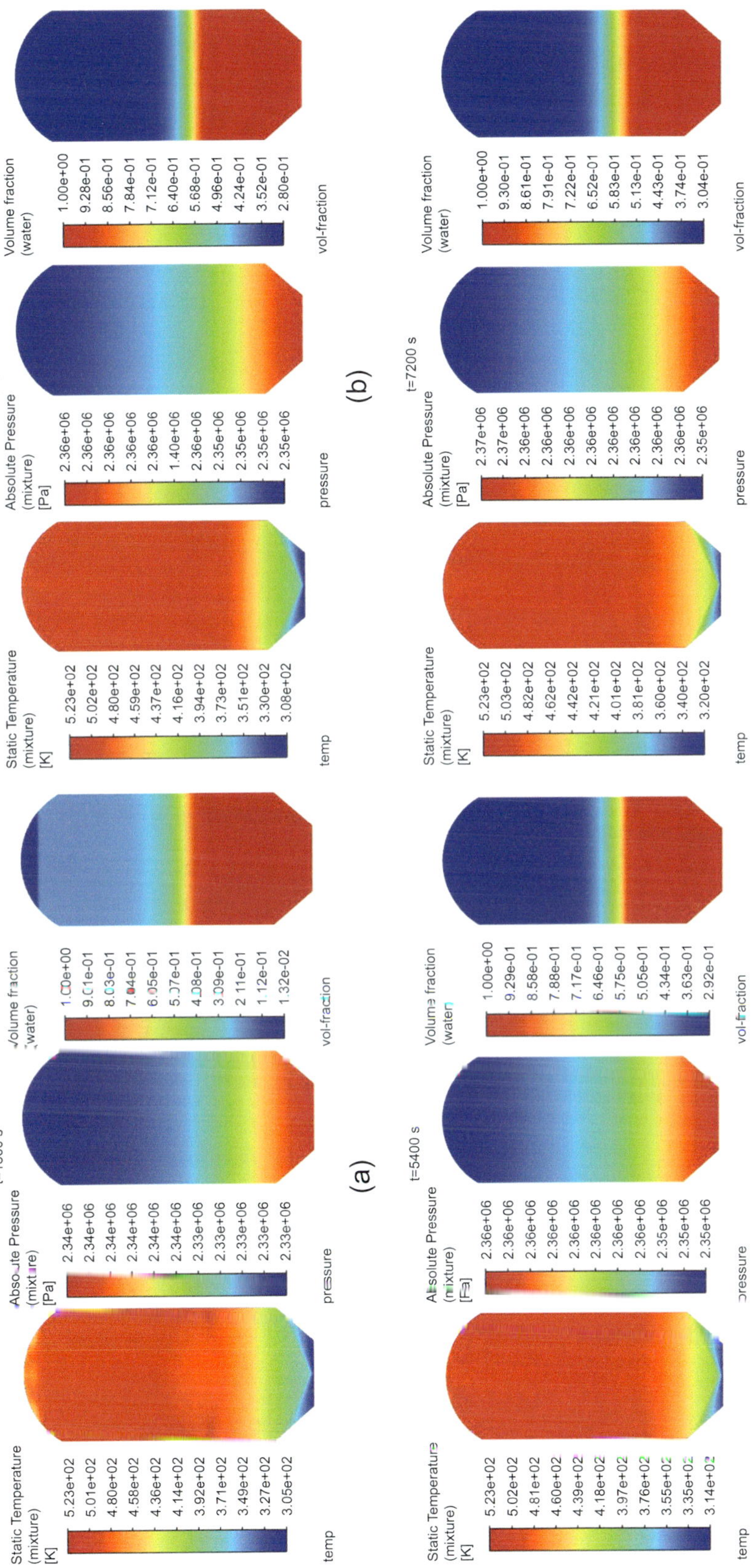

FIGURE 7.9 Contours of temperature, pressure, and water volume fraction at different flow times during the initial phase of the simulation: (a) $t = 1800$ s, (b) $t = 3600$ s, (c) $t = 5400$ s, and (d) $t = 7200$ s.

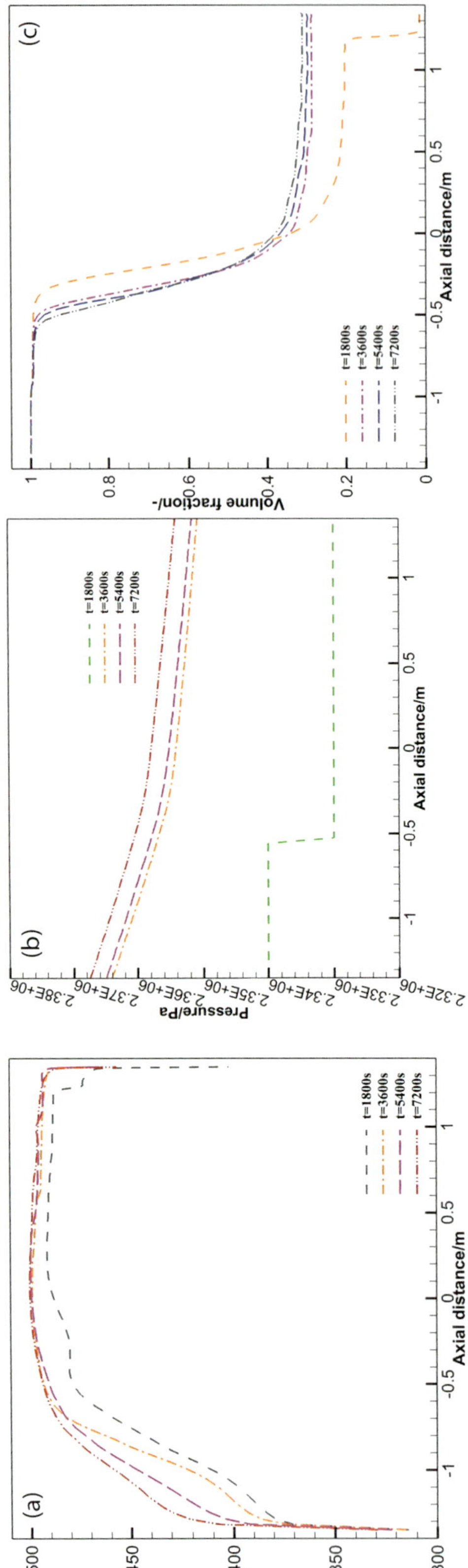

FIGURE 7.10 Axial flow profiles of the flow parameters at $y/d = 0.5$ towards quasi steady-state condition: (a) temperature, (b) pressure, and (c) volume fraction.

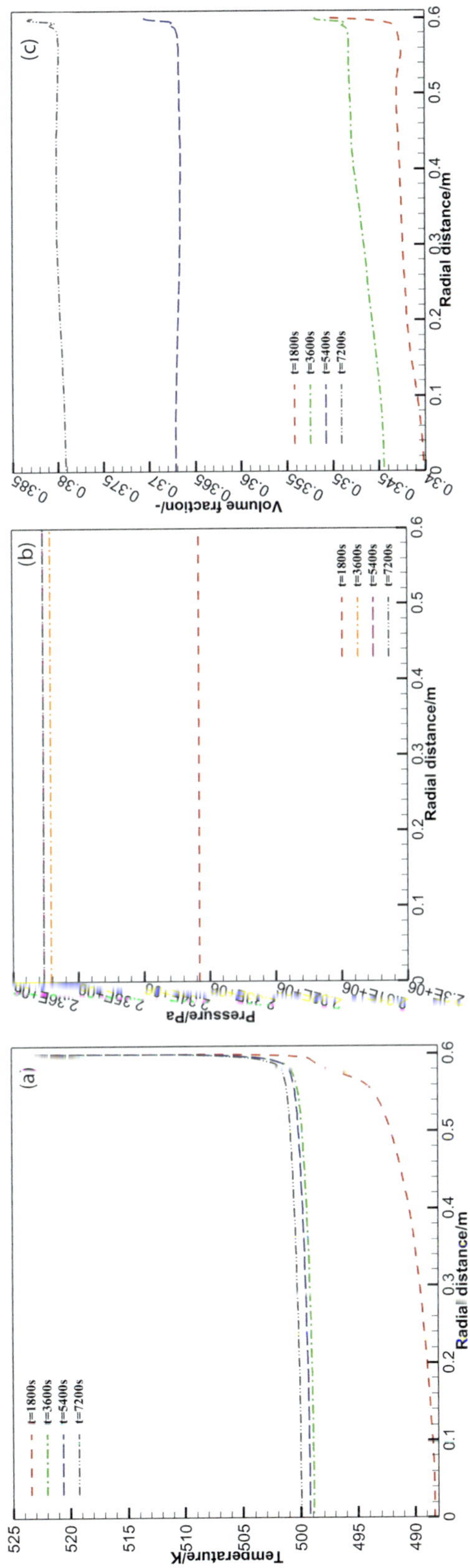

FIGURE 7.11 Radial flow profiles of the flow parameters at $x/H = 0.5$ towards quasi steady-state condition: (a) temperature, (b) pressure, and (c) volume fraction.

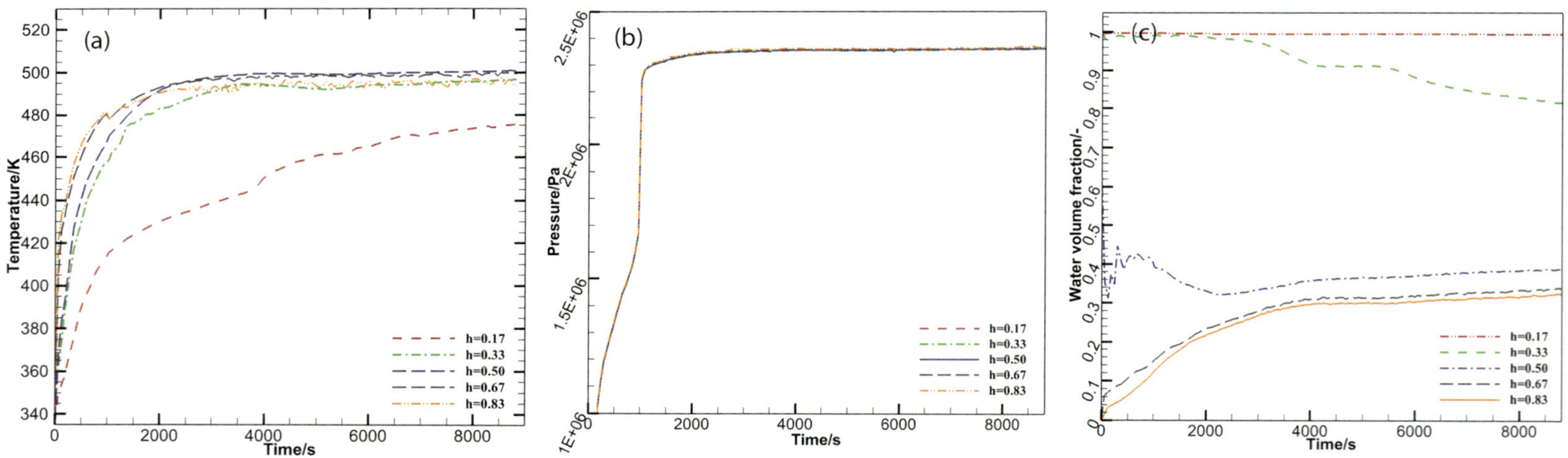

FIGURE 7.12 Spatial distribution of the flow parameters at different axial locations showing transient flow evolution towards steady-state condition: (a) temperature, (b) pressure, and (c) volume fraction.

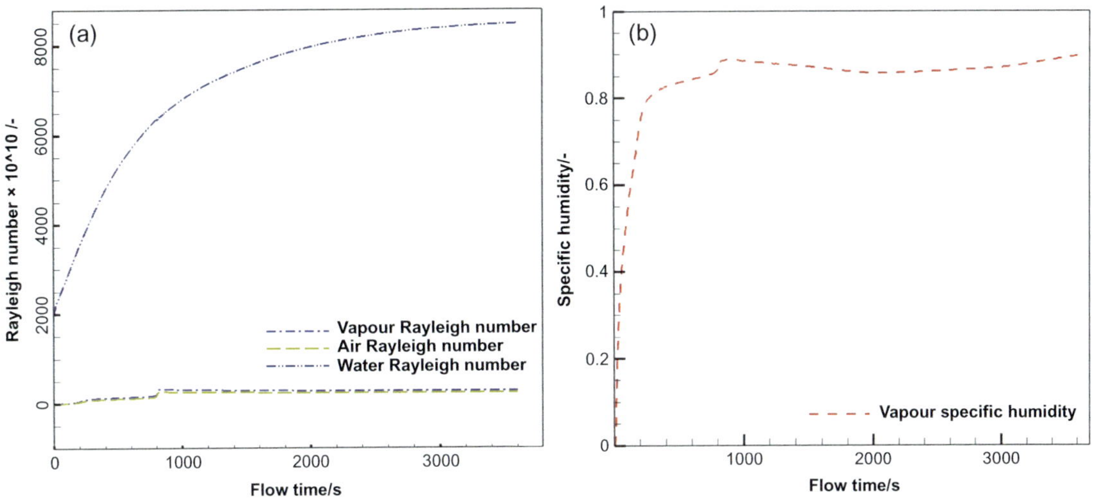

FIGURE 7.13 Transient behaviour of the fluid flow and thermodynamic properties: (a) Raleigh numbers as a function of time, (b) vapour specific humidity as a function of time.

Total heat energy supplied to the vessel Q_{wall} = Net heating power (W) × total flow time (s)
Total heat energy supplied to the vessel Q_{wall} = 97269.18 W × 9000 s = 875.40 MJ
Volume of water in the vessel V_w = volume of vessel/2 = 3.0 m^3/2 = 1.5 m^3

Initial mass of air (ideal gas) $m_a = \dfrac{P_a V M_a}{RT} = \dfrac{81375\,Pa \times 1.5\,m^3 \times 28.96\,gmol^{-1}}{8.3144\dfrac{J}{mol.K} \times 503.9K} = 0.87\,kg$

Mass of evaporated vapour (ideal gas) $m_v = \dfrac{P_v V M_w}{RT} = \dfrac{2797000\,Pa \times 1.5\,m^3 \times 18\,gmol^{-1}}{8.3144\dfrac{J}{mol.K} \times 503.9K} = 18.05\,kg$

P_v is the partial vapour pressure at 503.9 K taken from the steam table, M_w is the molar mass of water, T is the temperature, and R represents the gas constant.
Initial mass of water in the vessel before phase change $m_1 = \rho_w \times V_w$ = 923.9 kgm^{-3} × 1.5 m^3 = 1385.85 kg
Final mass of water in the vessel after phase change $m_2 = m_1 - m_v$ = (1385.85 kg − 18.05 kg) = 1367.80 kg

Change in internal energy of air:

$$\Delta U_a = m_a C_{va} \Delta T = 0.87\,kg \times \frac{0.7192KJ}{Kg\ K} \times \left(503.9K - 333K\right) = 0.107\,MJ$$

Change in internal energy of liquid water:

$$\Delta U_l = \Delta U_w + \Delta U_{vap} = \left(m_2 U_{l2} - m_1 U_{l1}\right) + \left(m_1 - m_2\right) U_g$$

Change in internal energy of liquid water:

$$\Delta U_w + \Delta U_{vap} = \left((1367.80kg * \frac{859.4KJ}{kg}) - (1385.85kg * \frac{251.2KJ}{kg}) \right)$$
$$+ \left(1385.85 - 1367.80\right) kg * 2602KJ / kg) = \left(1175.54 - 348.04\right) MJ + \left(46.97\right) MJ = 874.47MJ$$

859.4 kJ/kg and 251.2 kJ/kg represent the internal energy of water that exist in the liquid state inside the vessel at 475 K and 333 K, respectively, while 2602 kJ/kg represent the internal energy of the evaporated vapour at 503 K with all the values taken from the steam table.

It is worth pointing out that the actual net heat energy supplied to the vessel is slightly underpredicted by the solver as the heat flux varies with the flow time being highest at the beginning of the flow. However, the total change in the internal energy of the fluids derived from the first principles is very close to the predicted net heat energy from the Fluent solver. This solution check from the first principles gives a high degree of confidence in the predicted heat transfer parameters even in the absence of any experimental measurement of the heat transfer values.

7.5 CONCLUDING REMARKS

A VOF multiphase model has been developed to investigate fluid dynamics and heat transfer by natural convection in a pilot-scale HTC reactor as part of the commercial development of the reactor for sludge valorisation. Results of the investigation show the transient evolution of key process variables such as the temperature, water volume fractions, fluid flow regime, and thermodynamics of the process as well as the flow evolution towards the steady state. A steady-state condition was achieved in the reactor in approximately 150 min with a wall heat flux value of 15.109 kW/m^2 and a heat transfer rate of 97.269 kW. The predicted change in the internal energy by the generic solver is in agreement with the theoretical thermodynamic calculations. The mixing in the lower region of the reactor was characterised by thermal stratification and therefore needs some improvements. It is envisaged that in the subsequent stages of the project, investigation into alternative wall boundary condition (heat flux BC), as well as mixing performance and improvements will be conducted.

DATA AVAILABILITY

The data that support the findings of this chapter are available from the corresponding author, [B.O.], upon reasonable request.

NOMENCLATURE

Latin Letters

Variable	**Description, Units**
m	Phase mass transfer rate, N
Q_{wall}	Heat supplied by the heated wall, J
m_a	Initial mass of air, kg
m_1	Initial mass of water, kg
m_2	Final mass of water after evaporation, kg
m_v	Mass of evaporated vapour, kg
M_a	Molar mass of air, g/mol

M_w	Molar mass of water, g/mol
P_a	Partial pressure of air, Pa
P_v	Partial pressure of vapour, Pa
T	Temperature, K
R	Gas constant, J/mol K
ΔU_a	Change in internal energy of air, J
ΔU_w	Change in internal energy of water, J
ΔU_{vap}	Change in internal energy of vapour, J
ΔT	Change in temperature, K
c_{va}	Isochoric specific heat capacity of air, KJ/kg K
c_p	Isobaric specific heat capacity of fluid, J/kg K

Greek Symbols

Variable	Description, Units
ρ	Fluid density, kg/m^3
$\bar{v}$	Mean velocity, m/s
$\bar{p}$	Mean pressure, Pa
$\bar{\tau}_{ij}$	Viscous stress tensor, N/m^2
$\overline{\rho v_i' v_j'}$	Reynolds stress, N/m^2
$\vec{g}$	Gravitational body force, m/s^2
F_σ	Surface tension force, N
σ	Surface tension coefficient, -
γ	Mass transfer coefficient, -
μ	Fluid kinematic viscosity, kg/m s or (N s)/ m^2
T_{sat}	Saturation temperature, °C
P	Partial water pressure, Pa

Acronyms

HTC	Hydrothermal carbonisation
DS	Dry solids
HHV	High heating value
VOF	Volume of fluid
CAD	Computer-aided design
NIST	National Institute of Standards and Technology
HPC	High-performance computing
AD	Anaerobic digestion
NIST	National Institute of Standards and Technology

REFERENCES

1. G. Ischia, L. Fiori, Hydrothermal carbonization of organic waste and biomass: A review on process, reactor, and plant modeling, *Waste Biomass Valor.* 12 (2021) 2797–2824. https://doi.org/10.1007/s12649-020-01255-3.

2. M.P. Maniscalco, M. Volpe, A. Messineo, Hydrothermal carbonization as a valuable tool for energy and environmental applications: A review, *Energies* 13 (2020) 4098. https://doi.org/10.3390/en13164098.
3. A. Mahmoud, J. Olivier, J. Vaxelaire, A.F.A. Hoadley, Advances in mechanical dewatering of wastewater sludge treatment, in: S.K. Sharma, R. Sanghi (Eds.), *Wastewater Reuse and Management*, Springer, Heidelberg, 2013: pp. 253–303. https://doi.org/10.1007/978-94-007-4942-9_9.
4. A. Funke, F. Ziegler, Hydrothermal carbonization of biomass: A summary and discussion of chemical mechanisms for process engineering, *Biofuels Bioprod. Bioref.* 4 (2010) 160–177. https://doi.org/10.1002/bbb.198.
5. A. Hämäläinen, M. Kokko, V. Kinnunen, T. Hilli, J. Rintala, Hydrothermal carbonisation of mechanically dewatered digested sewage sludge—Energy and nutrient recovery in centralised biogas plant, *Water Res.* 201 (2021). https://doi.org/10.1016/j.watres.2021.117284.
6. K. Zhao, Y. Li, Y. Zhou, W. Guo, H. Jiang, Q. Xu, Characterization of hydrothermal carbonization products (hydrochars and spent liquor) and their biomethane production performance, *Bioresour. Technol.* 267 (2018) 9–16. https://doi.org/10.1016/j.biortech.2018.07.006.
7. W. Wang, W.-H. Chen, M.-F. Jang, Characterization of hydrochar produced by hydrothermal carbonization of organic sludge, *Future Cities Environ.* 6 (2020) 13. https://doi.org/10.5334/fce.102.
8. S. Román, J. Libra, N. Berge, E. Sabio, K. Ro, L. Li, B. Ledesma, A. Álvarez, S. Bae, Hydrothermal carbonization: Modeling, final properties design and applications: A review, *Energies* 11 (2018) 216. https://doi.org/10.3390/en11010216.
9. B. Mendecka, G. Di Ilio, L. Lombardi, Thermo-fluid dynamic and kinetic modeling of hydrothermal carbonization of olive pomace in a batch reactor, *Energies* 13 (2020) 4142. https://doi.org/10.3390/en13164142.
10. D. Knežević, W.P.M. Van Swaaij, S.R.A. Kersten, Hydrothermal conversion of biomass: I, glucose conversion in hot compressed water, *Ind. Eng. Chem. Res.* 48 (2009) 4731–4743. https://doi.org/10.1021/ie801387v.
11. H. Chater, M. Asbik, A. Koukouch, A. Mouaky, S. Bostyn, B. Sarh, F. Tabet, Analysis of fluid flow and heat transfer inside a batch reactor for hydrothermal carbonization process of a biomass, *Energies* 15 (2022) 818. https://doi.org/10.3390/en15030818.
12. L. Ding, W. Wang, B. Peng, B. Li, B. Li, B. Sunden, A modified dynamic Lee model for two-phase closed thermosyphon (TPCT) simulation, *Numer. Heat Transf. Part B: Fundam.* 85 (2024) 1041–1055. https://doi.org/10.1080/10407790.2023.2262114.
13. C.W. Hirt, B.D. Nichols, Volume of Fluid (VOF) method for the dynamics of free boundaries, *J. Comput. Phys.* 39 (1981) 201–225. https://doi.org/10.1016/0021-9991(81)90145-5.
14. J.U. Brackbill, D.B. Kothe, C. Zemach, A continuum method for modeling surface tension, *J. Comput. Phys.* 100 (1992) 335–354. https://doi.org/10.1016/0021-9991(92)90240-Y.
15. Z. Tan, Z. Cao, W. Chu, Q. Wang, Improvement on evaporation-condensation prediction of Lee model via a temperature deviation based dynamic correction on evaporation coefficient, *Case Stud. Therm. Eng.* 48 (2023) 103147. https://doi.org/10.1016/j.csite.2023.103147.
16. F. Marchelli, L. Fiori, Mixing due to natural convection in a hydrothermal reactor: A CFD study, *Ind. Eng. Chem. Res.* 63 (2024) 4134–4149. https://doi.org/10.1021/acs.iecr.3c03765.
17. D. Sangare, S. Bostyn, M. Moscosa-Santillan, I. Gökalp, Hydrodynamics, heat transfer and kinetics reaction of CFD modeling of a batch stirred reactor under hydrothermal carbonization conditions, *Energy* 219 (2021). https://doi.org/10.1016/j.energy.2020.119635.
18. K.S. Ro, J.A. Libra, A. Alvarez-Murillo, Comparative studies on water- and vapor-based hydrothermal carbonization: Process analysis, *Energies* 13 (2020) 5733. https://doi.org/10.3390/en13215733.

8 Mathematical Modelling of the Effects of Compression of Fouling Layers as a Predictive Tool in Membrane Bioreactor Systems

Suzanne Yala

8.1 INTRODUCTION

In recent years, mathematical modelling has attracted a lot of attention as a powerful tool that represents and describes the behaviour of a process [1]. This is due to the fact that the models are useful in evaluating and investigating complex systems with fewer or no experiments at all. This shortens the evaluation time, especially in processes that would require longer time to get experimental data [2, 3]. In addition, validated mathematical models can be exploited to predict the performance of a process under varied conditions and consequently use the results to design and optimise processes prior to applying them in a system. This minimises resources that are to be used by optimising the performance of the system.

During the modelling process, two key aspects need to be considered. First, the model representing a process needs to be built from realistic behaviour and laboratory data. Second, the developed model has to be validated using real experimental data. Generally, developing a new mathematical model involves key major steps [3, 4]:

DOI: 10.1201/9781003518426-8

i) Firstly, the objective of the modelling study needs to be defined. This is to avoid having a complicated model with regard to the set goals or an oversimplistic model that cannot answer the raised questions.
ii) Secondly, an investigation into prior knowledge of the system needs to be carried out. This includes knowledge on existing systems, models, and also historical, new, and literature data on a similar system.
iii) Thirdly, the type of model framework needs to be determined. That is, the type of model (mechanistic, empirical, and semi-empirical), the type of equation (algebraic, ordinary, or partial differential equation), and system boundary conditions.
iv) The fourth step is dependent on whether several models are available. If more than one model is available, then a model selection technique will have to be applied. Otherwise, this step can be skipped.
v) After the selection of a model, the next step is estimating the model parameters using experimental data.
vi) Lastly, the model is validated using experimental data from various conditions to test the accuracy of its predictability. If the validation test is successful, then the model can be used on other similar systems. Otherwise, the whole step of model development is repeated.

Fluid–particle separation is one of the areas that has widely adapted mathematical modelling for prediction, analysis, and minimisation of operational problems [1, 5, 6]. Hence, this chapter presents a case study of where mathematical modelling was employed in a fluid–particle separation system. The focus of the case study is on a membrane bioreactor (MBR) system, which consists of a physical fluid-separation process through a membrane and a biological process. The MBR technology is a well-established alternative process in wastewater treatment. However, it is still faced with the challenge of membrane fouling that results from the deposition and compression of particles on a membrane. This case study seeks to develop a model that can be used to get a prior prediction and understanding of the fouling problem and its effect on the membrane filtration performance. The study illustrates all the general steps that are involved in the development process of a mathematical model.

8.2 MODEL DEVELOPMENT

8.2.1 Description of the Problem

Wastewater of known composition (with s as the solid fraction of the suspension) is filtered through a membrane at a given flow rate (J) or applied pressure (ΔP). The membrane separates the liquid and particles by allowing the flow of liquid through it but retaining solids. Consequently, a fouling layer is formed on the membrane surface and grows in thickness with time, $L(t)$, as filtration proceeds. A schematic diagram of this process is presented in Figure 8.1.

During the filtration process, the liquid flowing through the formed fouling layer imparts drag force on the particles of the fouling layer, giving rise to a solid compressive pressure, P_s, which compacts the deposited foulants. Due to compression, the fouling layer structure undergoes a continuous change that varies with position and time. The properties affected are the local solidosity (ε_s), the local permeability (k), and the specific cake resistance (α). The change in these properties results in an increase in fouling resistance and a decline in permeate flux for constant pressure operations or an increase in total pressure drop for constant flux operations. This chapter seeks to describe the modelling process of the phenomenon of fluid flow through both the membrane and the fouling layer and the changes that come with it in terms of compression of the fouling layer in a membrane bioreactor (MBR).

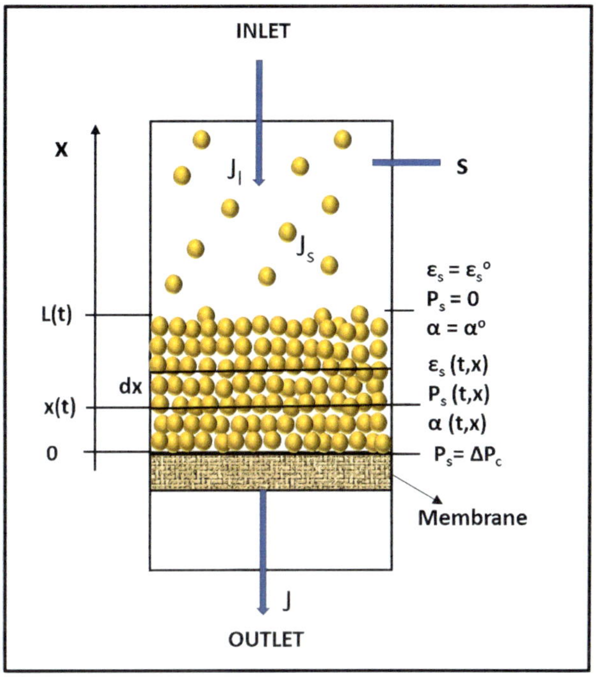

FIGURE 8.1 Schematic diagram of a dead-end membrane filtration model.

8.2.2 Governing Equations of Membrane Filtration of Compressible Foulant

The model developed in this chapter is based on conventional filtration theory for compressible material and was derived under the following assumptions:

i) The filtration process is in a dead-end mode. This is because this study is focused only on the effect of permeation drag as a result of fluid flow.
ii) The major fouling mechanism is cake fouling. This is in line with the literature, where cake layer formation is considered the largest contributor of fouling resistance in MBR systems [3–5].
iii) The fouling layer is treated as compressible. Studies have shown that fouling layers in MBR systems are largely compressible [6, 7]. Hence, it is important that this compressibility is taken into account in the modelling process.
iv) The fouling layer characteristics are a function of the local solid compressive stress [8–10].

Conventional filtration theory is a comprehensive modelling approach that gives a description of liquid flow through a compressible porous material with equations derived from Darcy's equation, momentum balance, and constitutive equation, which relate local cake properties (such as solidosity, permeability, and specific cake resistance) to solid compressive pressure.

In the cake filtration process, the fluid flow through a porous material such as a membrane for a one-dimensional case (see Figure 8.1) can be represented by Darcy's law in the form [10, 11]:

$$J = J_l 0 e J_s = \frac{k}{\mu}\frac{\partial P_L}{\partial x} \tag{8.1}$$

where J is the permeate velocity (m³/m²s or m/s), J_l and J_s denote the liquid and solid velocities (m/s), e is the void ratio, k is the cake permeability (m²), P_l is the pore liquid pressure (Pa), μ is the dynamic viscosity (Pa.s), and ∂x is the differential fouling layer thickness (m).

According to conventional cake filtration theory, J_s is always assumed negligible, hence Equation (8.1) is rewritten as

$$J = \frac{k}{\mu}\frac{\partial P_l}{\partial x} \tag{8.2}$$

As filtration occurs through a membrane at a flow rate, J, the amount of foulant deposited on the membrane increases. This in turn increases the thickness of the cake layer (L) with time, t. The growth rate of the fouling layer thickness ($\frac{dL}{dt}$) can be represented by Equations (8.3) and (8.4) [12, 13]:

$$\frac{dL}{dt} = \frac{\varepsilon_s^o}{\varepsilon_s^o - s}\left[\frac{k}{\mu}\frac{\partial P_l}{\partial x}\right]_{x=L} - J_{lm} \tag{8.3}$$

and

$$J_{lm} = \left[\frac{k}{\mu}\frac{\partial P_l}{\partial x}\right]_{x=0} \tag{8.4}$$

where ε_s^o is the fouling layer solidosity at zero solid compressive stress, s is the solid fraction in the suspension, P_l is the pore liquid pressure (Pa), μ is the dynamic viscosity (Pa.s), ∂x is the differential fouling layer thickness (m), and J_{lm} is the flow rate/flux of the membrane (m³/m²s).

As filtration progresses, fluid flowing through the formed fouling layer will exert a permeate drag force. The drag will force particles into existing voids within the cake, thus decreasing the porosity of the fouling layer [14]. This process is referred to as hydraulic compression. The extent of hydraulic compression at any plane in the fouling layer is determined by the effective compressive force, F_s, experienced by particles in that plane [15]. The compressive force exerted on a fouling layer plane of area A is known as solid compressive pressure, P_s [15]. This is expressed as follows:

$$P_s = \frac{F_s}{A} \tag{8.5}$$

The permeate drag forces generated by fluid flow through the fouling layer manifest as a decrease in the liquid hydraulic pressure, P_L, in the direction of flow. These drag forces are transmitted from particle to particle and eventually to the filter medium. At equilibrium, the effective compressive force, F_s, experienced by particles at a plane in the fouling layer is equal to the cumulative drag force up to that plane, as illustrated in Figure 8.2 [16], i.e.,

$$F_s = \sum F_d \tag{8.6}$$

where F_d is the fluid drag force on particles (N).

Also, at any plane in the cake, the cumulative drag forces are equal to the hydraulic pressure drop up to that point [17]:

$$\sum F_d = \left(P - P_L\right) \times A \tag{8.7}$$

where P is the pressure at the fouling layer surface (Pa) and P_L is the liquid hydraulic pressure (Pa).

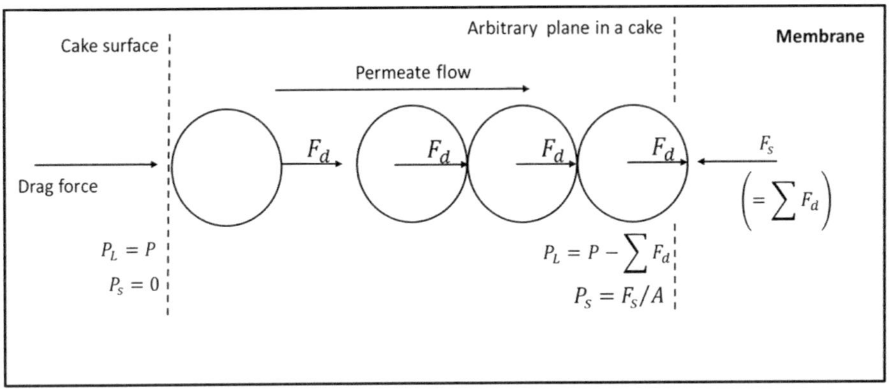

FIGURE 8.2 Representation of cumulative drag force through the fouling layer [15].

Combining Equations (8.5), (8.6), and (8.7) results in the momentum balance expression shown in Equation (8.8) [18]:

$$P_s + P_L = P \tag{8.8}$$

Differentiating Equation (8.8) across the whole cake thickness yields the following set of equations [19, 20]:

$$\frac{dP_s}{dL} + \frac{dP_L}{dL} = 0 \tag{8.9}$$

or

$$dP_s = -dP_l \tag{8.10}$$

On account of our previously stated assumption (see assumption (iv)), the local characteristics of compressible fouling layer in an MBR system are a function of solid compressive pressure. This relationship is commonly described by the constitutive expressions proposed by Tiller and Leu as presented in Equations (8.11)–(8.13) [10, 21, 22]:

$$\varepsilon_s = \varepsilon_s^o \left(1 + \frac{P_s}{P_a}\right)^{\beta} \tag{8.11}$$

$$k = k^o \left(1 + \frac{P_s}{P_a}\right)^{-\delta} \tag{8.12}$$

$$\alpha = \alpha^o \left(1 + \frac{P_s}{P_a}\right)^{n} \tag{8.13}$$

where ε_s, k, and $\alpha = \dfrac{1}{(\varepsilon_s k \rho_s)}$ are the cake solidosity, permeability (m^2), and specific cake resistance (m/kg), respectively. ε_s^o, k^o, and α^o denote the cake solidosity, permeability (m^2), and specific cake resistance (m/kg) at zero solid compressive pressure. P_s is the solid compressive pressure (Pa). P_a, β, $\delta = n + \beta$, and n are empirical constants, with the last three parameters signifying compression effects due to the solid compressive pressure. ρ_s is the density of the foulant.

Rearranging Equation (8.2) and using Equations (8.10) and (8.12) results in

$$\mu J \partial x = -k^{o}\left(1+\frac{P_s}{P_a}\right)^{-\delta} \partial P_s \tag{8.14}$$

Integrating Equation (8.14) from a fouling layer position $x = 0$ to $x = x$ (shown in Figure 8.1) gives

$$\mu \int_0^x J \partial x = \int_{\Delta P_c}^{P_s} k^{o}\left(1+\frac{P_s}{P_a}\right)^{-\delta} \partial P_s$$

$$\mu \int_0^x J \partial x = \frac{k^{o}}{1-\delta} P_a \left[\left(1+\frac{\Delta P_c}{P_a}\right)^{1-\delta} - \left(1+\frac{P_s}{P_a}\right)^{1-\delta}\right] \tag{8.15}$$

Similarly, integrating Equation (8.14) from cake position $x = 0$ (fouling layer/membrane interface) to $x = L$ (fouling layer surface) gives

$$\mu \int_0^L J \partial x = \int_{\Delta P_c}^{0} k^{o}\left(1+\frac{P_s}{P_a}\right)^{-\delta} \partial P_s$$

$$\mu \int_0^L J \partial x = \frac{k^{o}}{1-\delta} P_a \left[\left(1+\frac{\Delta P_c}{P_a}\right)^{1-\delta} - 1\right] \tag{8.16}$$

Dividing Equations (8.15) and (8.16) with the assumption that flux (J) remains constant across the fouling layer at time, t, we obtain an expression for the relative fouling layer thickness ($\frac{x}{L}$):

$$\frac{x}{L} = 1 - \frac{\left(1+\frac{P_s}{P_a}\right)^{1-\delta} - 1}{\left(1+\frac{\Delta P_c}{P_a}\right)^{1-\delta} - 1} \tag{8.17}$$

Rearranging Equation (8.17) yields an equation that can be used to approximate the solid compressive pressure profile through the fouling layer:

$$\frac{P_s}{P_a} = \left\{1 + \left(1 - \frac{x}{L}\right)\left[\left(1+\frac{\Delta P_c}{P_a}\right)^{1-\delta} - 1\right]\right\}^{1/(1-\delta)} - 1 \tag{8.18}$$

The liquid compressive profile through the cake can be obtained by combining Equation (8.18) with Equation (8.8) as follows:

$$\frac{P_l}{P_a} = \frac{P}{P_a} - \left\{1 + \left(1 - \frac{x}{L}\right)\left[\left(1+\frac{\Delta P_c}{P_a}\right)^{1-\delta} - 1\right]\right\}^{1/(1-\delta)} + 1 \tag{8.19}$$

Equations (8.18) and (8.19) describe the interaction between fluid and foulants in a membrane filtration process of compressible foulants.

Differentiating P_l in Equation (8.19) with respect to x gives the following equations:

$$\frac{\partial P_l}{\partial x} = \frac{P_a}{1-\delta}\frac{B}{L}\left[1+B\left(1-\frac{x}{L}\right)\right]^{\delta/1-\delta} \tag{8.20}$$

and

$$B = \left(1+\frac{\Delta P_c}{P_a}\right)^{1-\delta} - 1 \tag{8.21}$$

Combining Equations (8.4), (8.12), (8.18), and (8.20) with Equation (8.3) yields

$$\frac{dL}{dt} = \frac{s}{\varepsilon_s^o - s}\frac{k^o}{\mu}\frac{P_a}{1-\delta}\frac{B}{L} \tag{8.22}$$

Similarly, integrating Equation (8.22) from time $t = 0$ to $t = t$ gives an expression for estimating the fouling layer thickness formed during a particular filtration period:

$$L = \sqrt{\frac{2s}{\varepsilon_s^o - s}\frac{k^o}{\mu}\frac{P_a}{1-\delta}\int_0^t \left[\left(1+\frac{\Delta P_c}{P_a}\right)^{1-\delta} - 1\right] dt} \tag{8.23}$$

Substituting Equations (8.18) into (8.12) and then combining it with Equation (8.20) into Equation (8.2) yields the following results:

$$J = \frac{k^o}{\mu}\frac{P_a}{1-\delta}\frac{B}{L} \tag{8.24}$$

Combining Equations (8.21), (8.23), and (8.24) gives

$$J = \frac{\left(1+\frac{\Delta P_c}{P_a}\right)^{1-n-\beta} - 1}{\sqrt{\frac{2\mu s(1-n-\beta)}{k^o P_a\left(\varepsilon_s^o - s\right)}\int_0^t \left[\left(1+\frac{\Delta P_c}{P_a}\right)^{1-n-\beta} - 1\right] dt}} \tag{8.25}$$

From preliminary experimental results, it was found that the parameter P_a , if regarded as a constant as in previous studies, gave a poor fit between the model and experimental observations (see Section 8.5.1). This study subsequently proposed that P_a should be a function of applied pressure and that it increases with the increase in applied pressure. Therefore, a correlation between applied pressure (ΔP) and the parameter P_a was proposed as shown in Equation (8.26):

$$P_a = \tau e^{\Delta P \theta} \tag{8.26}$$

where τ and θ are empirical parameters associated with the feed.

Incorporating Equation (8.26) into Equation (8.25), we have

$$J = \frac{\left(1+\frac{\Delta P_c}{P_a}\right)^{1-n-\beta} - 1}{\sqrt{\frac{2\mu s(1-n-\beta)}{k^o \tau e^{\Delta P \theta}\left(\varepsilon_s^o - s\right)}\int_0^t \left[\left(1+\frac{\Delta P_c}{P_a}\right)^{1-n-\beta} - 1\right] dt}} \tag{8.27}$$

If the fouling layer characteristics $n, \beta, \tau, \theta, \varepsilon_s^o$ and k^o and the properties of the suspension being filtered such as s are known, Equation (8.27) can be used to calculate permeate flux over a certain filtration duration for a constant pressure membrane bioreactor system. Equation (8.27) relates the characteristics of the cake fouling layer, the solid compressive pressure, and the performance of an MBR (in terms of J vs t) for a constant pressure system. Therefore, it can be used to predict the performance of a membrane bioreactor under different compression conditions.

The performance of the membrane bioreactor can also be represented as total permeate volume collected versus time. The total permeate volume (V) versus time (t) can be calculated by substituting model data (J vs t) into Equation (8.28):

$$V = A\int_0^t J dt \tag{8.28}$$

By rearranging Equation (8.25), the pressure drop across the fouling layer at a time t_i for a constant flux system can be expressed as

$$\Delta P_c = P_a \left[\left[\frac{2J^2 \mu s(1-n-\beta)}{k^o P_a (\varepsilon_s^o - s)} t_i + 1 \right]^{1/(1-n-\beta)} - 1 \right] \tag{8.29}$$

Similarly, based on Darcy's equation, ΔP_c can be expressed as

$$\Delta P_c = \Delta P - \Delta P_m = \Delta P - \mu R_m J \tag{8.30}$$

where ΔP_c is the pressure drop across the fouling layer (Pa), ΔP_m is the pressure drop across the membrane (Pa), μ is the dynamic viscosity (Pa.s), R_m is the membrane resistance (m^{-1}), and J is the permeate flux ($m^3/(m^2s)$).

Finally, combining Equations (8.29) and (8.30) results in an equation that can be used to approximate the transmembrane pressure across the membrane at a time t_i during membrane filtration under a constant flux and this is presented in Equation (8.31):

$$\Delta P_i = \left(P_a \left[\left[\frac{2J^2 \mu s(1-n-\beta)}{k^o P_a (\varepsilon_s^o - s)} t_i + 1 \right]^{1/(1-n-\beta)} - 1 \right] + \Delta P_{i\text{-}1} \right) + \mu R_m J \tag{8.31}$$

8.3 MODEL SOLUTION ALGORITHM

The sequence of solving the proposed model equations for predicting flux decline (or total pressure drop), profiles of solid compressive pressure, porosity, and specific cake resistance is illustrated in Figure 8.3. The solution was implemented in MATLAB® R2025b (MathWorks Inc.), and the scripts are presented in the Appendix, Section A.8.2. Alternatively, the solution can also be implemented in Python 3.13.7 or newer versions using the same workflow but different Python scripts, which are presented as an Appendix in Section A.8.3.

The solution process is initiated by providing the values of model parameters $n, \beta, \tau, \theta, \varepsilon_s^o$ and k^o, the feed solid concentration, the membrane resistance, and the applied pressure (or flux). The procedure of

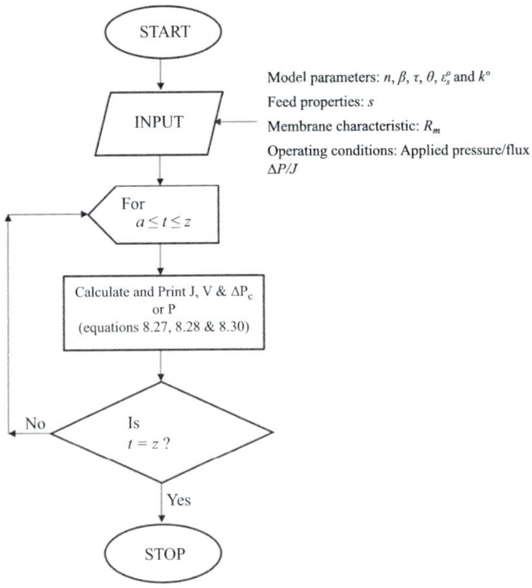

FIGURE 8.3 A flow chart illustrating the procedure of calculating permeate flux decline or total pressure drop using the proposed model equation in the membrane filtration of compressible particles.

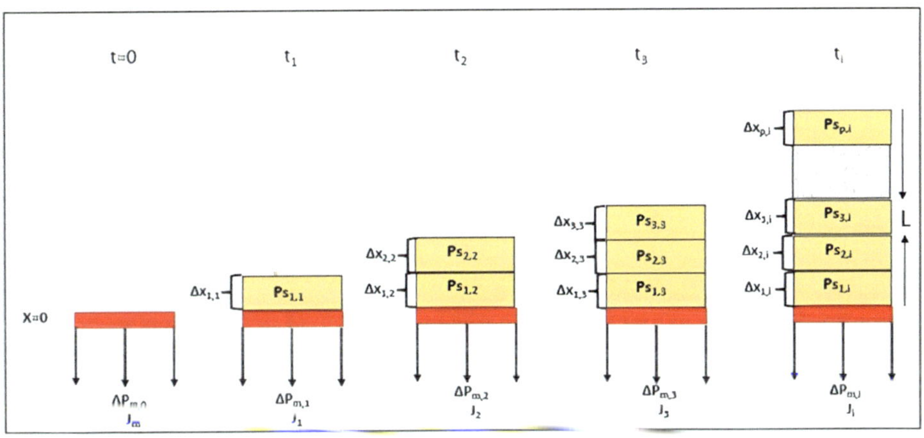

FIGURE 8.4 Illustration of fouling layer formation during the membrane filtration process.

obtaining the model parameters is outlined in the subsequent section. After these inputs, the calculation sequence proceeds. At a time, $t = 0$, no fouling layer has been formed and the only permeate flux (J_m) or total pressure drop present is that of the membrane. After a short time of t_1, a fouling layer x_1, as illustrated in Figure 8.4, is formed. The permeate flux (J_1) is now decreased, and the total pressure drop (ΔP) is divided between the membrane and the first fouling layer. The resultant permeate flux and the corresponding pressure drop across the layer (ΔP_{c1}) are obtained by solving Equations (8.27) and (8.30) simultaneously.

After a time t_2, a second layer x_2 is formed. The permeate flux decreases further to J_2, while the pressure drop across the cake increases to ΔP_{c2}. This is because the newly formed cake compresses the first layer, thus the compressive pressure within it increases and the porosity decreases, which decrease the permeate flux. The decrease in porosity also results in the increase in specific cake resistance that causes the increase in pressure drop across the cake. It is worth noting that the new layer is not compressed; hence its solid compressive pressure is zero and its porosity is denoted as ε^o. The sequence will therefore calculate all the parameters at t_2 and proceed to the next time interval.

A third fouling layer is then formed at time t_3. The permeate flux decreases further and the pressure drop across the layer increases. The first and second layers are compressed by the third layer. Hence, the solid compressive pressure in these two layers increases, while their porosity decreases. Again, the new layer is not compressed, its porosity will be ε^o. This process repeats itself until the filtration process ends. Similarly, the sequence of calculation is repeated until the end of the defined time interval. From the calculation results, a plot of permeate flux or total pressure drop vs time, at a defined time interval, is generated.

8.4 MODEL EVALUATION

8.4.1 Estimation of Model Parameters

The model parameters $n, \beta, \tau, \theta, \varepsilon_s^o$ and k^o were estimated by fitting the experimental data of permeate flux and transmembrane pressure versus time to the proposed Equations (8.27) and (8.31). The least-square method was employed for the estimation of parameters. The method involved minimising the objective function in Equation (8.32) by adjusting the parameters such that the model predictions fit the experimental data. This procedure was done in MATLAB R2025b (MathWorks Inc.) using the *fminsearch* function. The estimation of model parameters can also be implemented in Python 3.13.7 (see Appendix, Sections A.8.2 and A.8.3, respectively).

$$O\left(n,\beta,Pa,k^o,\varepsilon_s^o\right)=\sum_{i=1}^{N} r_i \tag{8.32}$$

where $r_i = \left(J_{i,\exp} - J_{i,\text{model}}\right)^2$ or $r_i = \left(P_{i,\exp} - P_{i,\text{model}}\right)^2$

and $J_{i,\exp}$ is the permeate flux measured experimentally corresponding to time t_i, $J_{i,\text{model}}$ is the value calculated from the developed model equations at the equivalent time t_i, and N denotes the total of J vs t pairs over which the number of analyses is performed. $P_{i,\exp}$ represents the total pressure drop data obtained from experimental run at time t_i and $P_{i,\text{model}}$ is the total pressure drop value calculated from the developed model equation corresponding to time t_i. The experimental data, used in the estimation of model parameters, were obtained from membrane filtration of 4 g/L of yeast suspension at 40 kPa for constant pressure operation and 35 LMH for constant flux operation. The filtration procedure is outlined in Section 8.4.3.

8.4.2 Sensitivity Analysis

Given that a large number of parameters were being estimated, it was important to assess the extent to which each parameter affects the prediction of the model. The analysis helped identify the most and the least sensitive model parameters. In the estimation process, the highly sensitive parameters will always need to be accurately calibrated; otherwise, a slight change in the parameter can result in huge errors [23]. On the other hand, the insensitive parameters will just need to be within the right range. This study used a one-factor-at-a-time (OFAT) strategy to perform the sensitivity analysis. The strategy involved changing the base values one at a time by ±15%, and the prediction of the effect of the change on the model was evaluated in terms of permeate. The ±15% was chosen as a typical example. The effects of the change of the base values of the parameters by ±15% on the model prediction were demonstrated using a 90-min 40 kPa constant pressure run as an example. Using the different parameter values, the permeate flux decline and thereafter the permeate volume were obtained using model Equations (8.27) and (8.28), respectively. The outcome was then used to calculate the corresponding percentage change in the total permeate volume collected.

8.4.3 Model Validation

After the calibration process, the predictive power of the model was validated. Model parameter values obtained in Section 8.4.1 by fitting the experimental data for 40 kPa constant pressure run and 35 LMH constant flux run were used. The validation process involved comparing experimental data sets that were not used in the estimation of the model parameters with the corresponding model predictions. The experimental data sets used included results obtained from filtration runs carried out at different pressure values (10, 20, 50, and 60 kPa), different fluxes (15, 25, 45, and 55 LMH), and with feed of different mixed liquor suspended solids (MLSS) (6 and 8 g/L). The filtration procedure under which the experimental data were obtained is outlined in Section 4.3. The model predictions were obtained by solving the proposed model equations in MATLAB R2025b (MathWorks Inc.) or Python 3.13.7 using the sample scripts (see Appendix, Sections A.8.2 and A.8.3, respectively). The model parameter values that were estimated in the previous section were used in the solution process. For constant pressure operations, the permeate flux decline was predicted by solving Equations (8.27) and (8.30) simultaneously. In constant flux operations, the change in total pressure drop was predicted by solving model Equation (8.31).

In the validation of the model, the comparison of the model predictions and the experimental data was assessed using the F test. The F test is a statistical tool that evaluates the equality of variance between two different data groups [24, 25]. The evaluation is always done by testing a null hypothesis and an alternative one. The null hypothesis suggests that the data sets are equal in variance and the obtained F value is less than the critical F value (i.e., is the upper quantile of the Fisher distribution evaluated with the different degrees of freedom). The alternative hypothesis shows otherwise. The obtained F value is calculated using Equation (8.33):

$$F_{val} = S_1^2 / S_2^2 \tag{8.33}$$

where $S_a = \dfrac{\Sigma(x_i - \bar{x})^2}{n-1}$

where S_1 is the variance of the first data set, which is always the larger value, and S_2 is the variance of the second data set. S_a is the variance of data set a which can either be the first or second data set, x_i is the data point i in a particular data set, $\bar{x}$ is the mean of that whole data set, and n is the number of data point. The $n-1$ expression represents the degree of freedom of that data.

In this study, the F test involves comparing the variance in experimental data to that predicted by the model for a particular run. The variance of each data set (S_a) was calculated from the data at each time interval (x_i). The analysis was performed in Excel using the two-sample F test function. The model was considered accurate if the F_{val} was relatively smaller than the critical F value for each run.

8.5 EXPERIMENTAL SECTION

8.5.1 Experimental Set-Up

The filtration experiments were performed using a laboratory-immersed membrane microfiltration unit, as shown in Fgure 8.5.

The rig consisted of a 2-L Plexiglas holding tank with a height of 33 cm, a flat sheet membrane module with an effective filtration area of 0.025 m^2, a pressure sensor of range –5 to –100 kPa for measuring pressure drop across the membrane, and a peristaltic pump for withdrawing permeate. Both the pressure sensor and the pump were obtained from Catchpoint Engineering & Allied, South Africa. The set-up also

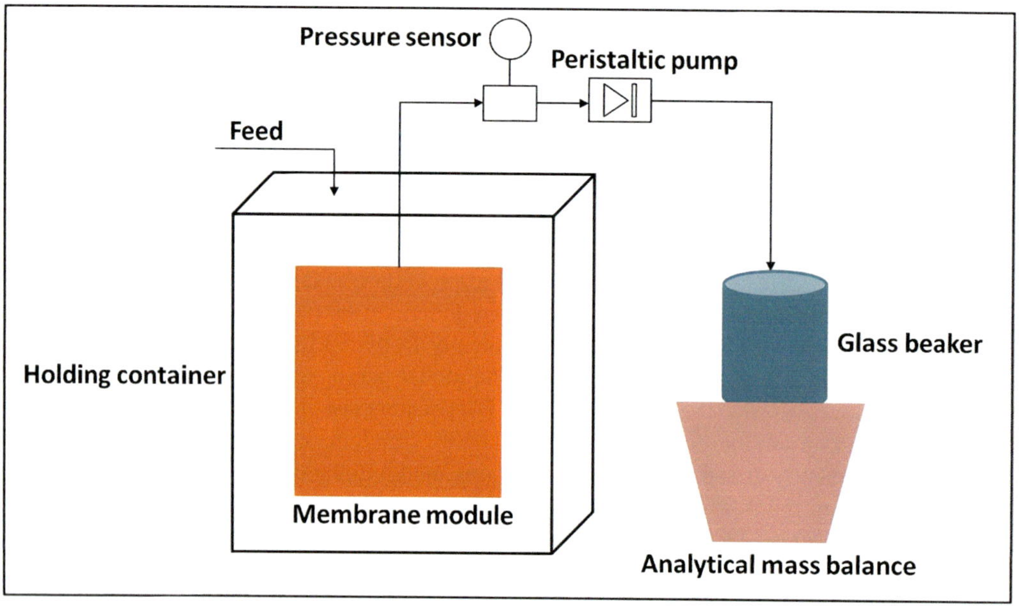

FIGURE 8.5 Schematic diagram of the laboratory-immersed membrane microfiltration unit.

consisted of a WBZ 15 analytical mass balance from Adam Equipment Co. Ltd, UK, for measuring the permeate mass. The membrane used in this work was an 800 kDa polyvinylidene fluoride (PVDF) microfiltration membrane bought from Synder Filtration, USA.

8.5.2 Filtered Suspension

In this case study, Baker's yeast was used as the model foulant. This is because it has been widely used in membrane fouling experiments, especially in investigations that involve the evaluation of compressible foulants [26–28]. In addition, this investigation wanted a reproducible feed that would make it easier to validate the model. Hence, yeast suspension (which is a synthetic wastewater feed) was chosen over real activated sludge for the testing phase of the model.

A yeast suspension of 4 g/L was employed in most of the filtration experiments, unless stated otherwise. The suspension was made by adding 4 g of Anchor Instant dry yeast to 1 L of distilled water. The content was then thoroughly mixed using a mixer. Fresh suspension was made for each investigation.

8.5.3 Filtration Experiment

The filtration experiments were carried out using the rig described in Section 8.4.1. Before each filtration experiment, distilled water was filtered through the membrane and the pure water flux was measured. These data were then used to calculate the intrinsic membrane resistance. The pure water flux runs were followed by a 90-min dead-end filtration experiment at either constant pressure or constant flux. During the filtration process, the change in permeate mass (for constant pressure runs) or total pressure drop (for constant flux runs) was recorded at an interval of 5 min throughout the experimental run. The experiments in this study were done at pressure and flux values that are close to the typical operating ranges of most MBR systems, that is, 10–50 kPa for constant pressure [29] and 15–55 L/m^{2}h (LMH) [30–32]. Each experimental run was repeated three times in order to ensure repeatability.

The recorded values of the change in permeate mass and the pressure drop were used to calculate permeate fluxes and fouling resistances using Equations (8.34) and (8.35), respectively:

$$J = \frac{\Delta m}{\rho A \Delta t} \tag{8.34}$$

where J is the permeate flux (m^3/m^2s), Δm is the mass of permeate collected (kg) during a time period of Δt (s), ρ is the density of water (kg/m^3), and A is the effective filtration membrane area (m^2):

$$R_f = \frac{\Delta P}{\mu J} - R_m \tag{8.35}$$

where R_f is the fouling resistance (m^{-1}), ΔP applied pressure (Pa), μ is the dynamic viscosity (Pa.s), J is the permeate flux (m^3/m^2s), and R_m is the intrinsic membrane resistance (m^{-1}).

8.6 RESULTS AND DISCUSSION

8.6.1 Preliminary Evaluation and Modification of the Original Model

The model developed in this study introduced two new parameters $\bar{\tau}$ and θ to account for the change in the value of empirical constant P_a with the change in applied pressure. In the preliminary evaluation of the initial model (Equation 8.27) derived from convention filtration theory, it was observed that the predictions fit well with data from some runs but were not in good agreement with data from other constant pressure runs (see Figure 8.6). For example, the prediction was consistent with experimental data for 40 kPa run, but the model overestimated the prediction of permeate flux decline for 10 kPa run and underestimated that of 60 kPa run. This showed that there was a fundamental flaw in the model derived from conventional cake filtration theory.

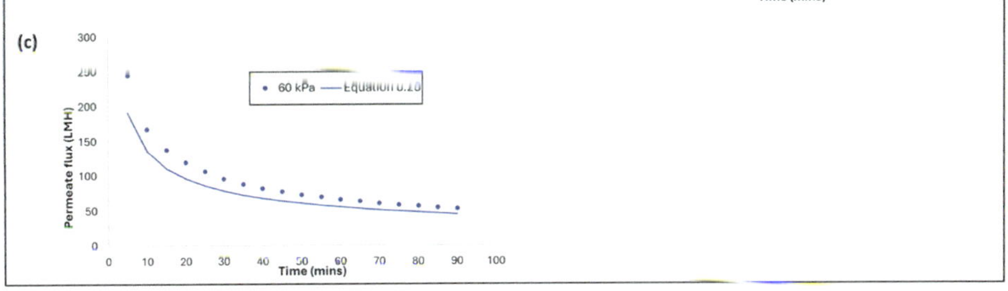

FIGURE 8.6 Comparison of experimental data and predictions by model Equation (8.25) of permeate flux decline for various constant pressure runs in membrane filtration of 4 g/L of yeast suspension.

After an intensive investigation of each model parameter, it was identified that the value of the parameter P_a was causing a discrepancy in the model prediction. Further evaluation showed that an adjustment of the value of P_a gave a better fit for the different constant pressure runs, and the P_a value increased with the increase in applied pressure. This finding deviates from most studies, which have always considered P_a to be only a function of feed and not applied pressure. However, a study by Sioutopoulos and Karabelas hinted that P_a may be affected by applied pressure. They reported that the value of P_a is also affected by the initial permeate flux [33]. Therefore, the wide variation in the initial permeate flux for the different constant pressure runs resulted in different P_a values. This study proposed to account for this variation by formulating a correlation between applied pressure and P_a (see Equation 8.26). The formulation was then incorporated into the proposed model Equation (8.27). The modified model equation gave better predictions that align with the experimental data as would be seen in the subsequent section.

It is worth noting that the modification did not apply to constant flux runs. The model predictions of total pressure drop for constant flux runs did fit well with experimental data, without the adjustment of the value of P_a for different applied flux. This may be associated with the small difference between the applied fluxes as compared to in constant pressure runs, where the initial fluxes differ by a wider margin.

8.6.2 Estimation of Model Parameters and Sensitivity Analysis

The values of model parameters, $n, \beta, \tau, \theta, \varepsilon_s^o, P_a$ and k^o were obtained by fitting the experimental data of 40 kPa run to Equation (8.27) and experimental data of 35 LMH run to Equation (8.31). This resulted in parameter values presented in Table 8.1, with a good agreement between the experimental data and the model prediction as shown in Figures 8.7 and 8.8.

To improve the parameter estimation process, a sensitivity analysis was carried out to identify parameters that have a high impact on model predictions and thus will need to be accurately calibrated. This is to avoid huge errors. Among the parameters listed in Table 8.1, the compressibility coefficient (n) was identified as the most sensitive parameter, while β was the least sensitive (see Figure 8.9).

The significant influence of the n parameter is attributed to the fact that it is the primary quantity that characterises the compressibility effect, which in turn has a strong influence on the performance of membrane filtration systems [27, 34]. A ±15% change in n resulted in an equal change in the cumulative volume of permeate collected, indicating its sensitive nature. On the other hand, the same variation in model parameter β produced only a ±1.5% change in the predicted permeate volume. The variation of the parameters $\varepsilon_s^o, k^o, \tau$ and θ by ±15% resulted in a change between ±6 and 8% of the permeate volume at the end of the run. Hence, during the curve fitting, a lot of attention was paid to the compressibility coefficient. The β parameter was not varied that much.

TABLE 8.1 Estimated Model Parameters of Yeast Suspension Determined by Equations (8.27) and (8.31) under 40 kPa Constant Pressure and 35 LMH Constant Flux, Respectively

PARAMETER	*SYMBOL*	*VALUES*
Compressibility coefficient	n	1.37 ± 0.07
Empirical constant	β	0.15
Empirical constant	τ	0.2877
Empirical constant	θ	0.022
Cake permeability at $P_s = 0$	$k^o (m^2)$	$(1.89 \pm 0.02) \times 10^{-14}$
Cake solidosity at $P_s = 0$	ε_s^0	0.27 ± 0.03
Empirical constant (constant flux operations)	P_a (Pa)	1300 ± 200

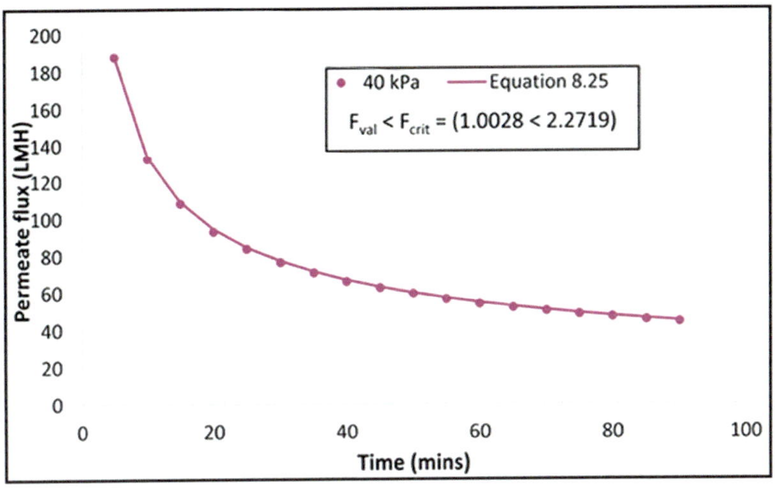

FIGURE 8.7 Comparison of experimental data and model prediction of permeate flux versus time for a 40 kPa constant pressure run in membrane filtration of 4 g/L of yeast suspension (F_{val} – *F* value and F_{crit} – critical *F* value).

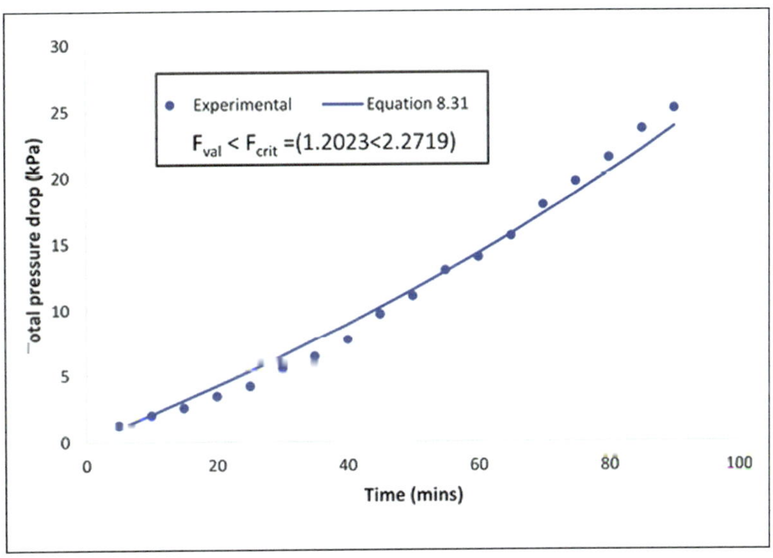

FIGURE 8.8 Comparison of experimental data and model prediction of total pressure drop versus time for a 35 LMH constant flux run in membrane filtration of 4 g/L of yeast suspension (F_{val} – *F* value and F_{crit} – critical *F* value).

8.6.3 Validation of the Model

The estimated model parameters were validated using data sets from other runs carried out at different applied pressures, at different fluxes, and with different feed concentrations that were not used in the estimation process.

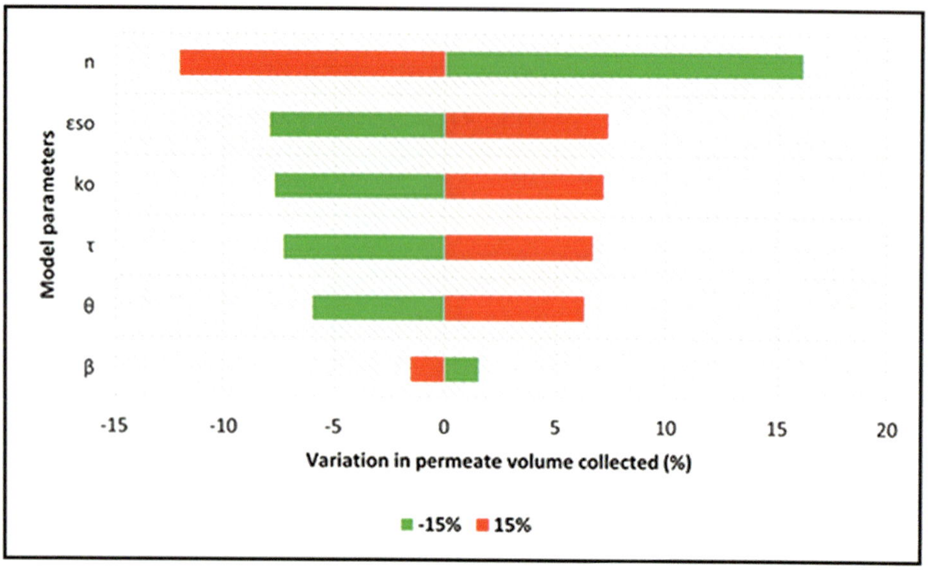

FIGURE 8.9 Sensitivity analysis of the effect of variation of the model parameters on the prediction of permeate volume collected at the end of a 90-min 40 kPa constant pressure run. A ±15% change in the parameter base values was applied.

8.6.3.1 Different Applied Pressures

A comparison of the experimental data and model predictions of permeate flux decline with time for various applied constant pressure runs is shown in Figure 8.10.

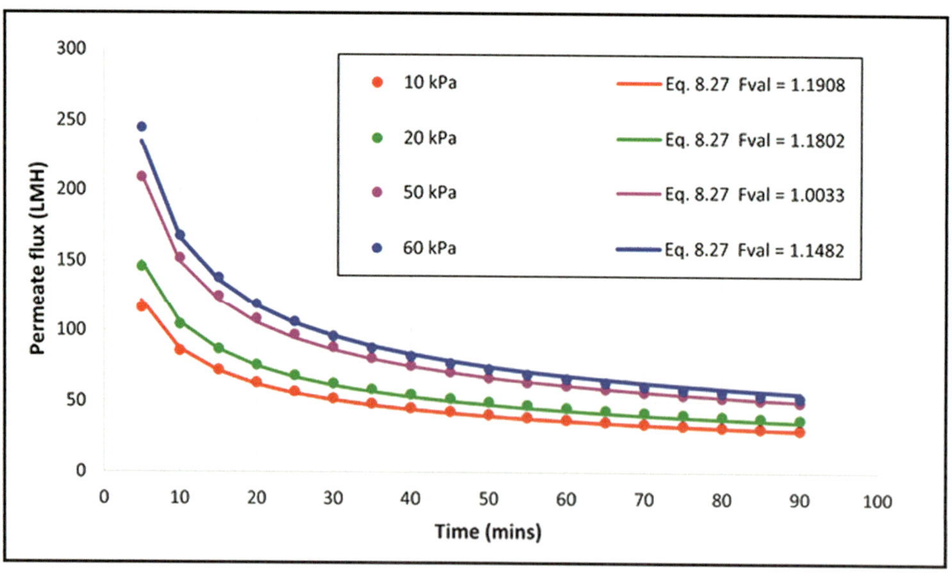

FIGURE 8.10 Comparison of experimental data and model prediction of permeate flux versus time for various constant pressure filtration runs (predictions were done using parameters obtained from fitting experimental data of 40 kPa run). The *F* critical value in the comparison was reported to be 2.2719.

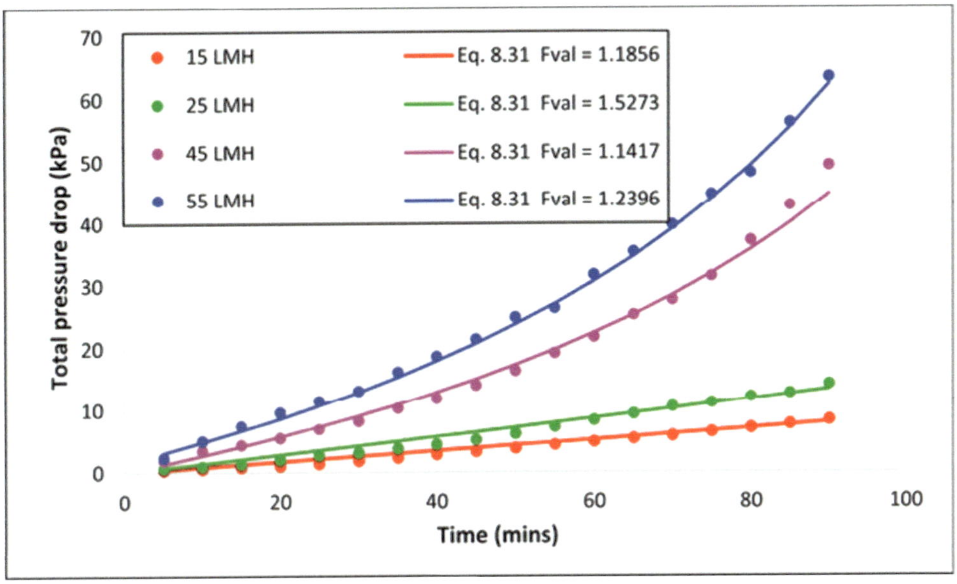

FIGURE 8.11 Comparison of experimental data and model prediction of transmembrane pressure versus time for various constant flux filtration runs (predictions were done using parameters obtained from fitting experimental data of 35 LMH run). The *F* critical value in the comparison was reported to be 2.2719.

8.6.3.2 Different Applied Fluxes

Similarly, there was consistency between model predictions and experimental data for constant flux runs, as shown in Figure 8.11. This is supported by the *F*-test data (see Figure 8.11), which indicates that the experimental data fit well with the proposed model for the different applied fluxes. The *F* values for all the runs fall between 1.00 and 1.26. These are less than the critical *F* value that is around 2.719.

On the basis of the presented data and the above observations, it can therefore be concluded that the proposed model equations can satisfactorily predict the permeate flux decline (for constant pressure operations) or the increase in pressure drop (for constant flux operations) in the membrane filtration of compressible foulants, based on the filtration and compression characteristics of the feed.

8.6.3.3 Different MLSS Concentrations

The proposed model was also validated with yeast suspensions of different concentrations, as shown in Figures 8.12 and 8.13. For the constant pressure runs carried out at 40 kPa, Figure 8.12. show that the experimental data fit well with the model predictions for yeast concentrations of 6 g/L and 8 g/L. The goodness of fit is further confirmed by the *F* test. The obtained *F* values for both feed concentrations are less than the *F* critical value, implying that the variance of the experimental data and the model prediction are statistically equal. The experimental data for constant flux operations also agreed well with model predictions as depicted in Figure 8.13.

The *F*-test results further confirm the prediction accuracy of the model. The *F* values obtained from comparing model predictions and experimental data for runs carried out with a yeast concentration of 6 g/L (1.0231) and 8 g/L (1.0239) were less than the critical *F* value (2.2719). Hence, the null hypothesis that the variances for both data sets were equal was true. This indicates that the proposed model can accurately predict the performance of membrane in the filtration of different feed concentrations under both constant pressure and constant flux operations.

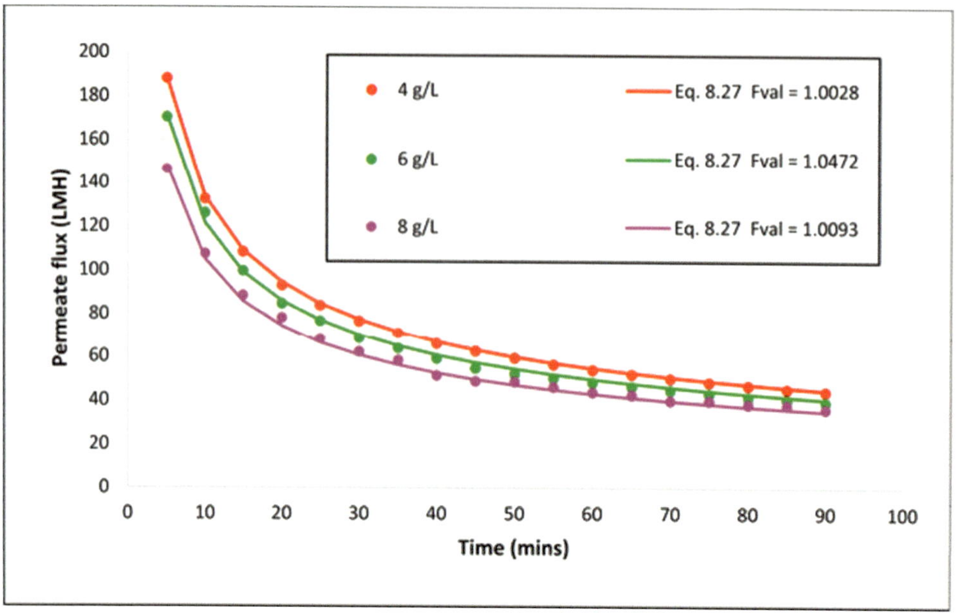

FIGURE 8.12 A comparison of model predictions and experimental data of permeate flux decline in the membrane filtration of yeast suspension of different concentrations at constant pressure of 40 kPa (predictions were done using parameters obtained from fitting experimental data of 4 g/L yeast suspension). The *F* critical value in the comparison was reported to be 2.2719.

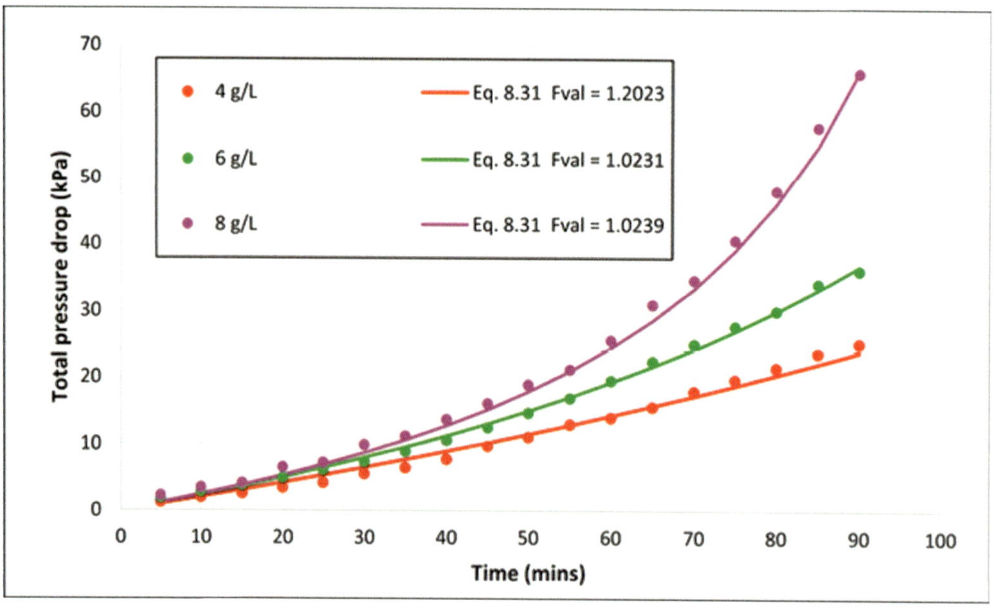

FIGURE 8.13 A comparison of model predictions and experimental data of the increase in transmembrane pressure in membrane filtration of yeast suspension of different concentrations at a constant flux of 35 LMH (predictions were done using parameters obtained from fitting experimental data of 4 g/L yeast suspension). The *F* critical value in the comparison was reported to be 2.2719.

8.6.4 Applicability of the Model to Real Wastewater

The ultimate objective of this study was to develop a model that can be used to predict the performance of a membrane bioreactor in the filtration of wastewater and also the effect of different compression conditions on the spatial characteristics of the fouling layer formed during the filtration process. Hence, it was essential to assess the applicability of the proposed model on real wastewater. In this work, the model was evaluated using municipal wastewater obtained from Stellenbosch Wastewater Works. The sample was collected from the influent of the MBR system and used within 8 h. The characteristic of the wastewater sample is presented in Table 8.2.

The model parameters associated with the municipal wastewater were obtained by fitting the experimental data of 40 kPa run into model Equation (8.27) and 60 LMH data into model Equation (8.31), as outlined in Section 3.1. The values are shown in Table 8.3. The fitted data and the corresponding model predictions are found in the Appendix, Figures A8.1 and A8.2. The different membrane filtration runs, in which the model parameters were obtained, were carried out using different batches of wastewater samples collected on different days. This was to ensure that the variability of wastewater characteristics, if any, is factored in the model parameter estimation. From the data in Table 8.3, it can be seen that there was a minimal variation in the values of different model parameters obtained using different wastewater batches. The variation was around or less than 10% for all the parameters. The minimal variation in the estimated values is attributed to the little variation in terms of the wastewater characteristics, as shown by experimental results of membrane performance in Figures A8.1 and A8.2 in the Appendix. In the figures, the minimal variation is depicted by the small error bars. It should be noted that the minimal variation in wastewater characteristics may not be observed from samples obtained in varied seasons or from other wastewater works, as it has been widely reported that real wastewater varies from day to day due to physio-biochemical processes [35–37]. In this study, the sampling was done within a period of two months, where a wide variation of the wastewater characteristics was not observed.

The obtained model parameters were then validated using experimental data from other constant pressure and constant flux runs. From Figure 8.14, it can be observed that the experimental data fit well into model Equation (8.29) for all the evaluated constant pressure runs. This is further supported by the

TABLE 8.2 Characteristics of the Municipal Wastewater Sample Collected from the Stellenbosch Municipal Wastewater Works

CHARACTERISTIC	*VALUE*
MLSS (g/L)	9–11
COD (g/L)	10–12
DO (mg/L)	2–3

TABLE 8.3 Estimated Model Parameters for Municipal Wastewater Sample Determined by Equations (8.27) and (8.31) under 40 kPa Constant Pressure and 35 LMH Constant Flux, Respectively

PARAMETER	*SYMBOL*	*VALUES*
Compressibility coefficient	n	1.1 ± 0.1
Empirical constant	β	0.1 ± 0.01
Empirical constant	τ	0.3877
Empirical constant	θ	0.031 ± 0.001
Cake permeability at $P_s = 0$	$k^0 (m^2)$	$(5.9 \pm 0.01) \times 10^{-14}$
Cake solidosity at $P_s = 0$	ε_s^0	0.44 ± 0.01
Empirical constant (constant flux operations)	P_a (Pa)	1050 ± 50

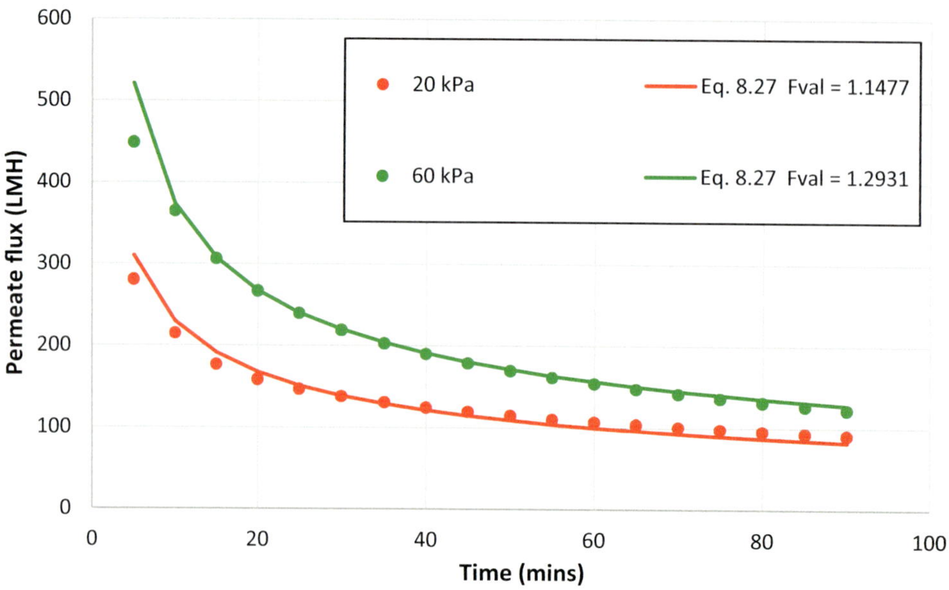

FIGURE 8.14 Comparison of model predictions and experimental data obtained from membrane filtration of 10 g/L of municipal wastewater under various constant pressure operations (predictions were done using parameters obtained from fitting experimental data of 40 kPa run). The *F* critical value in the comparison was reported to be 2.2719.

F-test results (see Figure 8.14). The *F* values obtained from the comparison of model predictions and the experimental data for the runs carried out at 20 and 60 kPa were all less than the *F* critical value (2.2719).

For constant flux operations, the comparison of model predictions and experimental data for the increase in pressure drop over time is presented in Figure 8.15. In this case too, the model predictions were

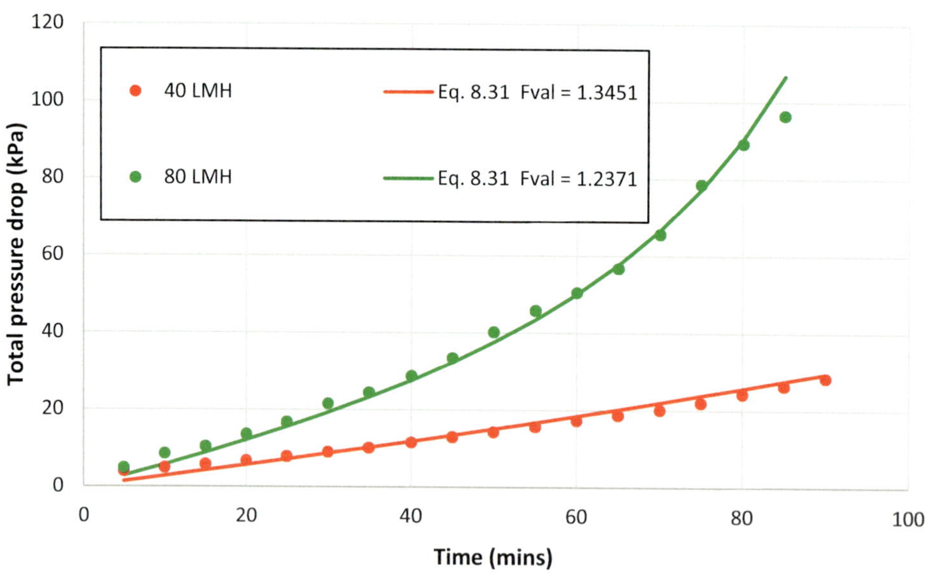

FIGURE 8.15 Comparison of model predictions and experimental data obtained from the membrane filtration of 10 g/L of municipal wastewater under various constant flux operations (predictions were done using parameters obtained from fitting experimental data of 60 LMH run). The *F* critical value in the comparison was reported to be 2.2719.

consistent with the experimental results for all the evaluated runs. This is also confirmed by the F values, which were less than the F critical value for runs done at 40 and 80 LMH (see Figure 8.15).

From the above observations and discussions, it can be concluded that the proposed model can be applied to real membrane bioreactor systems to predict the filtration performance of membranes in both constant pressure and constant flux operations.

8.7 CONCLUSIONS

This chapter describes a case study on mathematical modelling of an operational problem in a fluid–particle separation system. The case study is on modelling the effect of compression of fouling layers on membrane filtration performance. First, the development of mathematical model equations based on conventional cake filtration theory was outlined. The proposed model was intended for predicting flux decline or increase in the pressure drop in a membrane bioreactor, based on the filtration and compressibility characteristics of a feed. The model development process was followed by the estimation of model parameters through the fitting of experimental data obtained from the membrane filtration of yeast suspension. The proposed model Equations (8.27) and (8.31) were then validated with runs carried out under different operating conditions and feed concentration. In all the evaluated conditions, the predictions of the model were found to be in good agreement with the experimental data. In addition, the model equations were also found to be reliable in predicting flux decline and an increase in transmembrane pressure in membrane filtration of real municipal wastewater. Therefore, the proposed model provides a valuable tool for predicting the filtration performance of the membrane bioreactor and can be used to gain information on operational problems that affect the performance of fluid–particle separation processes in membrane systems.

DATA AVAILABILITY

The data that support the findings of this chapter are available from the corresponding author, [S.Y.], upon reasonable request.

NOMENCLATURE

Variable	Description, Units
A	Effective membrane area, m^2
e	Void ratio, —
F_d	Fluid drag force, N
F_s	Compressive force, N
F_{val}	F value in F test, —
F_{crit}	Critical F value in F test, —
J	Permeate flux, LMH
J_l	Liquid velocity, m/s
J_{lm}	Membrane flux, LMH
$J_{i, exp}$	Permeate flux measured experimentally, LMH
$J_{i, model}$	Permeate flux predicted by the model, LMH

k	Permeability, m^2
k^o	Initial permeability, m^2
L	Fouling layer thickness, m
Δm	Mass of permeate collected, kg
n	Compressibility coefficient, —
ΔP	Applied pressure, Pa
P	Total pressure drop across the membrane and the fouling layer, Pa
P_a	Empirical constant, Pa
ΔP_c	Pressure drop across the fouling layer, Pa
$P_{i, exp}$	Total pressure drop measured experimentally, Pa
$P_{i, model}$	Total pressure drop predicted by the model, Pa
P_l	Liquid hydraulic pressure, Pa
P_m	Pressure drop across the membrane, Pa
P_s	Solid compressive pressure, Pa
R_f	Fouling resistance, m^{-1}
R_m	Membrane resistance, m^{-1}
s	Solid fraction in suspension, —
S_1	Variance of the first data set, —
S_2	Variance of the second data set, —
Δt	Time taken for filtration, h
V	Total permeate volume, L
x	Fouling layer thickness, m

Greek Symbols

Variable	**Description, Units**
ε°	Initial porosity, —
ε	Porosity, —
ε_s^o	Initial solidosity, —
μ	Dynamic viscosity, Pa.s
α°	Initial specific cake resistance, m/kg
α	Specific cake resistance, m/kg
ρ	Feed density, kg/m^3
∂x	Differential fouling layer thickness, m
δ, β	Parameter signifying compressibility, —
τ, θ	Empirical constants, —

Acronyms

COD	Chemical oxygen demand
DO	Dissolved oxygen
LMH	Litre/metre^2-hour
MBR(s)	Membrane bioreactor(s)
MLSS	Mixed liquor-suspended solids
O	Objective function
RIS	Resistance in series

APPENDIX

A.8.1 Supplementary Experimental Charts

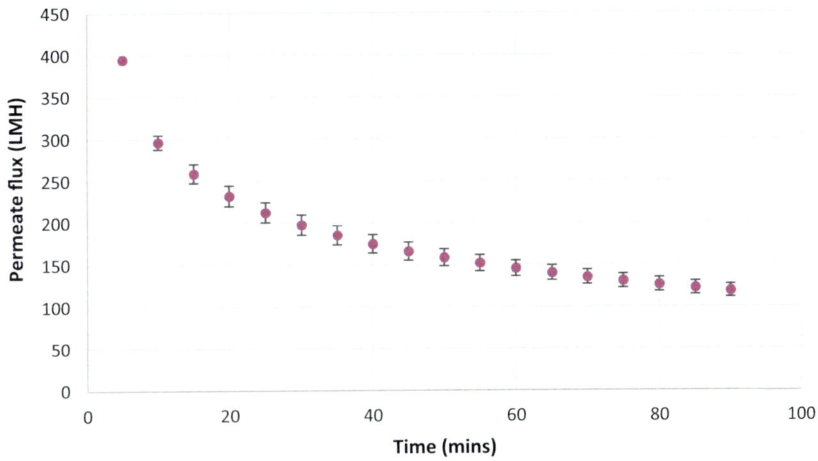

FIGURE A8.1 Experimental data of permeate flux decline of a 40 kPa constant pressure run in membrane filtration of 10 g/L of municipal wastewater.

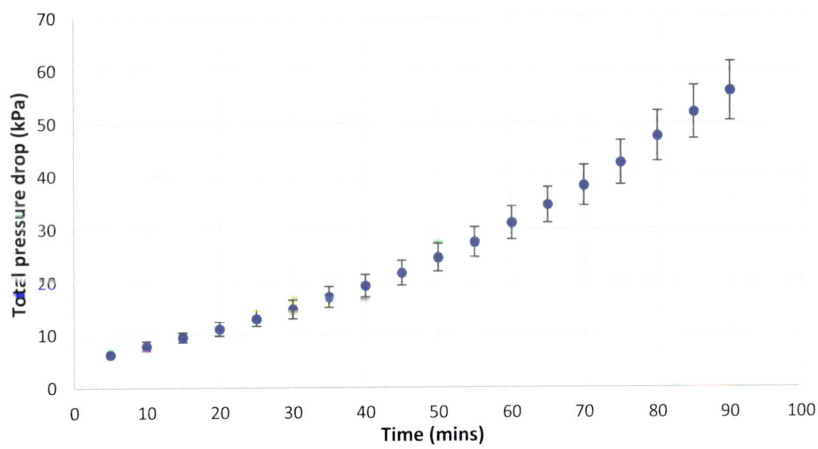

FIGURE A8.2 Experimental data of total pressure drop versus time for a 60 LMH constant flux run in membrane filtration of 10 g/L of municipal wastewater

A.8.2 Supplementary MATLAB Codes

i) Script for the estimation of model parameters

```
function Estimated_parameters ()
% Input experimental data (replace with actual values if needed)
Exp_t = [300, 600, 900, 1200, 1500, 1800, 2100, 2400, 2700, 3000, 3300, 3600,
3900, 4200, 4500, 4800, 5100, 5400];
```

```
Exp_J1 = [5.22222E-05, 3.67778E-05, 3.00741E-05, 2.58333E-05, 2.32444E-05,
2.12222E-05, 1.96825E-05, 1.83333E-05, 1.73827E-05, 1.64889E-05, 1.56566E-05,
1.4963E-05, 1.44103E-05, 1.38889E-05, 1.34074E-05, 0.000013, 1.25621E-05,
1.22469E-05];

Exp_DeltaPc = [33481.93333, 35416.98438, 36238.53236, 36781.70458,
37091.31275, 37352.03542, 37538.26589, 37713.58443, 37827.31111, 37938.66142,
38044.58, 38132.84549, 38201.26429, 38265.72869, 38327.02956, 38375.57557,
38432.78838, 38470.06491];

% Initial guess for model parameters
initial_bestx = [1.37, 0.15,0.3877,0.055, 1.89E-14, 0.3];

% Estimate model parameters using fminsearch
estimated_parameters = fminsearch(@(bestx) para_function(bestx, Exp_t, Exp_J1,
Exp_DeltaPc), initial_bestx);

% Display estimated parameters
fprintf('Estimated Parameters:\n');
fprintf('n = %.4f\n', estimated_parameters(1));
fprintf('beta = %.4f\n', estimated_parameters(2));
fprintf('tau = %.4f\n', estimated_parameters(3));
fprintf('theta = %.4f\n', estimated_parameters(4));
fprintf('k0 = %.4e\n', estimated_parameters(5));
fprintf('eps0 = %.4f\n', estimated_parameters(6));

% Plot the fitted curve vs experimental data
figure;
plot(Exp_t / 60, Exp_J1, 'ro', 'MarkerFaceColor', 'r'); % Experimental data
hold on;

y_fit = model_funct(estimated_parameters, Exp_DeltaPc, Exp_t); % Fitted model
plot(Exp_t/ 60, y_fit, 'b-', 'LineWidth', 2);
xlabel('Time (minutes)');
ylabel('Permeate flux (m3/m2.s)');
legend('Experimental Data', 'Model');
title('Data Fitting for J vs t');
grid on;

% Nested function to define the model
   function J_fitted = model_funct(bestx, Exp_DeltaPc, Exp_t)

      n = bestx(1);
      beta =(bestx(2));
      tau = bestx(3);
      theta = bestx(4);
      k0 = bestx(5);
      eps0 = bestx(6);

      s = 0.00398;
      mu = 0.00089;
      DeltaP=40000;

      A = 1 - n - beta;
      Pa=(tau * exp((DeltaP / 1000) * theta))* 1000;
      C = (2 * mu * s * A) / (k0 * Pa * (eps0 - s));
```

```
      J_fitted = -((((1 + (Exp_DeltaPc ./ Pa)).^A) - 1) ./ (sqrt(C * (((1 +
      (Exp_DeltaPc ./ Pa)).^A) - 1) .* Exp_t)));

   end
   % Nested function to define the cost function
   function para = para_function(bestx, Exp_t, Exp_J1, Exp_DeltaPc)
      J_model = model_funct(bestx, Exp_DeltaPc, Exp_t);
      para = sum((Exp_J1 - J_model).^2); % Sum of squared errors
   end
end
```

ii) Script for solving all the model equations in the mathematical model

```
% Define the Parameters (replace with actual values from experimental data
fitting)
clear;

n = 1.8213;               % Example value for n
beta = 0.1671;            % Example value for beta
tau = 0.3660;             % Example value for tau
theta = 0.0264;           % Example value for theta
epsilon_so = 0.3010;      % Example value for epsilon_so
k_o = 1.8739e-14;         % Example value for k_o
DeltaP = 50000;           % Example value for Delta P
mu = 0.00089;             % Example value for mu
Rm = 1.37e11;             % Example value for Rm
s=0.00398;                % Example value for s
rhos=1095.2;              % Example value for density of the feed

A=1-n-beta;
Pa=(tau*exp((DeltaP/1000)*theta))*1000;
C=(2*mu*s*A)/(k_o*Pa*(epsilon_so-s));
l=0;
y=zeros(54,2);

% Define the time interval and step
t_start = 100; % start time
t_end = 90*60; % end time in seconds (90 minutes)
dt = 100;        % time step in seconds

% Initialize variables
t_values = t_start:dt:t_end; % time values
J_values = zeros(size(t_values)); % initialize J values
DeltaPc_values = zeros(size(t_values)); % initialize DeltaPc values

for t=100:100:5400

      % Solve the equations simultaneously using fsolve
      % Initial guesses for [J, DeltaPc]
      x0 = [100 5*10^(-5)];

      % Define the system of equations as a function handle
      F = @(x) [ x(1) - DeltaP + (mu*Rm.*x(2));
                 x(2) + ( ((1 + (x(1)/Pa))^A - 1) ./ sqrt( C * (((1 + (x(1)/
                 Pa))^A) - 1) * t ) ) ];
```

```
        % Solve using fsolve
        m = fsolve(F, x0);

        % Extract solutions
        y(l+1,1)=m(1);
        y(l+1,2)=m(2);
        l=l+1;

        % Store the values
        J_values = y(:,2);
        DeltaPc_values= y(:,1);
end

%% Solve for Ps, alpha, and epsilon for all DeltaPc values
f_values = 0:0.01:1; % Values for f
Ps_values = zeros(length(f_values), length(DeltaPc_values));
alpha_values = zeros(length(f_values), length(DeltaPc_values));
epsilon_values = zeros(length(f_values), length(DeltaPc_values));

alpha_o = 1 / (epsilon_so * k_o * rhos);

% Iterate over each DeltaPc_value
for j = 1:length(DeltaPc_values)
    DeltaPc_value = DeltaPc_values(j); % Use the j-th calculated DeltaPc value
    V = (1 + DeltaPc_value / Pa)^A;

    % Iterate over f values
    for i = 1:length(f_values)
        f = f_values(i);

        % Solve for Ps
        Ps0 = 40000; % Initial guess for Ps
        F1 = @(Ps) (Ps / Pa) - (1 + (1 - f) * (V - 1))^(1 / A) + 1;
        Ps_values(i, j) = fsolve(F1, Ps0);

        % Solve for epsilon
        epsilon0 = 0.5; % Initial guess for epsilon
        F2 = @(epsilon) epsilon - 1 + epsilon_so * ((V - 1) * (1 - f) +
        1)^(beta / A);
        epsilon_values(i, j) = fsolve(F2, epsilon0);

        % Solve for alpha
        alpha0 = 4e13; % Initial guess for alpha
        F3 = @(alpha) alpha - alpha_o * ((V - 1) * (1 - f) + 1)^(n / A);
        alpha_values(i, j) = fsolve(F3, alpha0);
    end
end

figure (1); % Plot Permeate flux vs Time
plot(t_values / 60, J_values * 1000 * 3600); % Convert t_values to minutes
xlabel('Time (minutes)');
ylabel('Permeate flux (LMH)');
title('J vs Time');
grid on;
```

```
figure (2); % Plot solid compressive pressure profile across the fouling layer
hold on
for r = [1 54]
   plot(f_values,Ps_values(:,r))
   xlabel("Relative cake thickness, x/L")
   ylabel("Solid compressive pressure (kPa)")
   title('Ps vs Time')
   grid on;
end

figure (3); % Plot porosity profile across the cake layer
hold on
for r = [1 54]
   plot(f_values,epsilon_values(:,r))
   xlabel("Relative cake thickness, x/L")
   ylabel("Porosity")
   title('Porosity vs Time')
   grid on;
end
hold off
```

iii) Script for predicting flux decline in a constant pressure run

```
% Define the parameters (replace with actual values)
n = 1.37; % Example value for n
beta = 0.15; % Example value for beta
tau = 0.2877; % Example value for tau
theta = 0.022; % Example value for theta
epsilon_so = 0.27; % Example value for epsilon_so
k_o = 1.89*10^(-14); % Example value for k_o
DeltaP = 50000; % Example value for Delta P
mu = 0.00089; % Example value for mu
Rm = 1.37*10^11; % Example value for Rm
s=0.00398;

A=1-n-beta;
Pa=(tau*exp((DeltaP/1000)*theta))*1000;
C=(2*mu*s*A)/(k_o*Pa*(epsilon_so-s)),
n=0;
p=0;
y=zeros(54,2);

% Define the time interval and step
t_start = 100; % start time
t_end = 90*60; % end time in seconds (90 minutes)
dt = 100; % time step in seconds

% Initialize variables
t_values = t_start:dt:t_end; % time values
J_values = zeros(size(t_values)); % initialize J values
DeltaPc_values = zeros(size(t_values)); % initialize DeltaPc values
```

```
for t=100:100:5400

    % Solve the equations simultaneously using fsolve
    % Initial guesses for [J, DeltaPc]
    x0 = [100 5*10^(-5)];

    % Define the system of equations as a function handle
    F=@(x)[x(1)-DeltaP+(mu*Rm.*x(2)),x(2)+((((1+(x(1)/Pa))^A)-1)./(sqrt(C*
    (((1+(x(1)./Pa))^A)-1).*t)))];

    % Solve using fsolve
    m =fsolve(F,x0);

    % Extract solutions
    y(n+1,1)=m(1);
    y(n+1,2)=m(2);
    n=n+1; % Increment index

    % Store the values
    J_values = y(:,2);
    DeltaPc_values= y(:,1);
end

% Plot J vs. t
figure;
plot(t_values/60, J_values * 1000 * 3600); % t_values/60 to convert to
minutes
xlabel('Time (minutes)');
ylabel('Permeate flux (LMH)');
title('Permeate flux vs. Time');
grid on;
```

iv) Script for predicting the total pressure drop increase in a constant flux run

```
% Define the parameters (replace with actual values)
clear;
n = 1.35;
beta = 0.15;
Pa = 1100;
k0 = 1.89 * 10^(-14);
eps0 = 0.3; s = 0.00398;
mu = 0.00089;
J = 9.722 * 10^(-6);
Rm = 1.12 * 10^(11);

A = 1 - n - beta;
C = (2 * mu * s * A) / (k0 * Pa * (eps0 - s));

% Define the time interval and step
t_start = 300;          % start time
t_end = 90*60;          % end time in seconds (90 minutes)
dt = 300;               % time step in seconds
t_values = t_start:dt:t_end; % time values
t_values=t_values';
```

```
% Initialize cumulative P value
P_total = 0;
n = 0; % Loop index
y = zeros(18, 1);

for t = 300:300:5400
    % Define the function for fsolve
    F = @(P) (P - ((((((J^2) * C * t) + 1).^(1 / A)) - 1) * Pa) - (mu * Rm *
    J));

    % Initial guess for fsolve
    P0 = 100;

    % Solve for P
    P = fsolve(F, P0);

    % Accumulate P into P_total
    P_total = P_total + P;

    % Store current P and cumulative P_total if needed
    y(n + 1, 1) = P; % Current P at each time step
    y(n + 1, 2) = P_total; % Cumulative P value up to this step

    TotalP_values = y(:,2);
    n = n + 1; % Increment index
end

% Plot P_total vs. t
figure;
plot(t_values/60, TotalP_values/ 1000); % t_values/60 to convert to minutes
xlabel('Time (minutes)');
ylabel('Total pressure drop (kPa)');
title('Total pressure drop vs. Time');
```

A.8.3 Supplementary Python Codes

i) Script for the estimation of model parameters

```
# Import python packages and libraries
import numpy as np
from scipy.optimize import minimize
from scipy.optimize import fsolve
import matplotlib.pyplot as plt
from numba import njit, prange

def estimated_parameters():
    # Experimental data
    Exp_t = np.array([300, 600, 900, 1200, 1500, 1800, 2100, 2400,
                      2700, 3000, 3300, 3600, 3900, 4200, 4500,
                      4800, 5100, 5400])

    Exp_J1 = np.array([5.22222E-05, 3.67778E-05, 3.00741E-05, 2.58333E-05,
                       2.32444E-05, 2.12222E-05, 1.96825E-05, 1.83333E-05,
                       1.73827E-05, 1.64889E-05,
```

```
                        1.56566E-05, 1.4963E-05, 1.44103E-05, 1.38889E-05,
                        1.34074E-05, 0.000013, 1.25621E-05, 1.22469E-05])

    Exp_DeltaPc = np.array([33481.93333, 35416.98438, 36238.53236, 36781.70458,
                            37091.31275, 37352.03542, 37538.26589,
                            37713.58443, 37827.31111, 37938.66142,
                            38044.58, 38132.84549, 38201.26429, 38265.72869,
                            38327.02956, 38375.57557,
                            38432.78838, 38470.06491])

    # Initial guess
    initial_bestx = np.array([1.37, 0.15, 0.3877, 0.055, 1.89E-14, 0.3])
    # Optimization
    result = minimize(lambda bestx: para_function(bestx, Exp_t, Exp_J1,
                                                  Exp_DeltaPc), initial_bestx,
method='Nelder-Mead')
    est_params = result.x
    # Display results
    print("Estimated Parameters:")
    print(f"n = {est_params[0]:.4f}")
    print(f"beta = {est_params[1]:.4f}")
    print(f"tau = {est_params[2]:.4f}")
    print(f"theta = {est_params[3]:.4f}")
    print(f"k0 = {est_params[4]:.4e}")
    print(f"eps0 = {est_params[5]:.4f}")
    # Plot results
    plt.figure()
    plt.plot(Exp_t / 60, Exp_J1, 'ro', label='Experimental Data')
    y_fit = model_funct(est_params, Exp_DeltaPc, Exp_t)
    plt.plot(Exp_t / 60, y_fit, 'b-', linewidth=2, label='Model')
    plt.xlabel('Time (minutes)')
    plt.ylabel('Permeate flux (m3/m2.s)')
    plt.title('Data Fitting for J vs t')
    plt.legend()
    plt.grid(True)
    plt.show()

def model_funct(bestx, Exp_DeltaPc, Exp_t):
    n, beta, tau, theta, k0, eps0 = bestx
    s = 0.00398
    mu = 0.00089
    DeltaP = 40000
    A = 1 - n - beta
    Pa = (tau * np.exp((DeltaP / 1000) * theta)) * 1000
    C = (2 * mu * s * A) / (k0 * Pa * (eps0 - s))
    J_fitted = -(((1 + (Exp_DeltaPc / Pa)) ** A - 1) /
                 np.sqrt(C * ((1 + (Exp_DeltaPc / Pa)) ** A - 1) * Exp_t))
    return J_fitted

def para_function(bestx, Exp_t, Exp_J1, Exp_DeltaPc):
    J_model = model_funct(bestx, Exp_DeltaPc, Exp_t)
    return np.sum((Exp_J1 - J_model) ** 2) # Sum of squared errors

if __name__ == "__main__":
    estimated_parameters()
```

ii) Script for solving all the model equations in the mathematical model

```
# Define the Parameters (replace with actual values from experimental data
fitting)
n = 1.8213
beta = 0.1671
tau = 0.3660
theta = 0.0264
epsilon_so = 0.3010
k_o = 1.8739e-14
DeltaP = 50000
mu = 0.00089
Rm = 1.37e11
s = 0.00398
rhos = 1095.2
A = 1 - n - beta
Pa = (tau * np.exp((DeltaP / 1000) * theta)) * 1000
C = (2 * mu * s * A) / (k_o * Pa * (epsilon_so - s))
alpha_o = 1 / (epsilon_so * k_o * rhos)
# Time settings
t_start = 100
t_end = 90 * 60
dt = 100
t_values = np.arange(t_start, t_end + dt, dt)
num_steps = len(t_values)

# Preallocate
J_values = np.zeros(num_steps)
DeltaPc_values = np.zeros(num_steps)

# Solve for J and DeltaPc at each time step
x0 = [100, 5e-5] # Initial guess

for idx, t in enumerate(t_values):
    def F(x):
        return [
            x[0] - DeltaP + (mu * Rm * x[1]),
            x[1] + (((1 + (x[0] / Pa)) ** A - 1) / np.sqrt(C * ((1 + (x[0] /
Pa)) ** A - 1) * t))
        ]
    m, _, flag, _ = fsolve(F, x0, full_output=True)
    DeltaPc_values[idx] = m[0]
    J_values[idx] = m[1]
    x0 = m # Use previous solution as next initial guess

# f, Ps, alpha, epsilon calculations
f_values = np.arange(0, 1.01, 0.01)
num_f = len(f_values)
Ps_values = np.zeros((num_f, num_steps))
alpha_values = np.zeros((num_f, num_steps))
epsilon_values = np.zeros((num_f, num_steps))

for j in range(num_steps):
    DeltaPc = DeltaPc_values[j]
    V = (1 + DeltaPc / Pa) ** A
    for i, f in enumerate(f_values):
```

```
        # Ps
        Ps0 = 40000
        def F1(Ps):
            return (Ps / Pa) - (1 + (1 - f) * (V - 1)) ** (1 / A) + 1
        Ps_sol = fsolve(F1, Ps0)[0]
        Ps_values[i, j] = Ps_sol

        # epsilon
        epsilon0 = 0.5
        def F2(epsilon):
            return epsilon - 1 + epsilon_so * ((V - 1) * (1 - f) + 1) ** (beta
/ A)
        epsilon_sol = fsolve(F2, epsilon0)[0]
        epsilon_values[i, j] = epsilon_sol

        # alpha
        alpha0 = 4e13
        def F3(alpha):
            return alpha - alpha_o * ((V - 1) * (1 - f) + 1) ** (n / A)
        alpha_sol = fsolve(F3, alpha0)[0]
        alpha_values[i, j] = alpha_sol

# Plot Permeate flux vs Time
plt.figure(1)
plt.plot(t_values / 60, J_values * 1000 * 3600)
plt.xlabel('Time (minutes)')
plt.ylabel('Permeate flux (LMH)')
plt.title('J vs Time')
plt.grid(True)

# Plot solid compressive pressure profile across the fouling layer
plt.figure(2)
for r in range(num_steps):
    plt.plot(f_values, Ps_values[:, r])
plt.xlabel('Relative cake thickness, x/L')
plt.ylabel('Solid compressive pressure (kPa)')
plt.title('Ps vs Time')
plt.grid(True)

# Plot porosity profile across the cake layer
plt.figure(3)
for r in range(num_steps):
   plt.plot(f_values, epsilon_values[:, r])
plt.xlabel('Relative cake thickness, x/L')
plt.ylabel('Porosity')
plt.title('Porosity vs Time')
plt.grid(True)

plt.show()
```

iii) Script for predicting flux decline in a constant pressure run

```
# Import python packages and libraries
import numpy as np
from scipy.optimize import fsolve
import matplotlib.pyplot as plt
```

```
# Define the parameters (replace with actual values)
n_val = 1.37        # Example value for n
beta = 0.15         # Example value for beta
tau = 0.2877        # Example value for tau
theta = 0.022       # Example value for theta
epsilon_so = 0.27   # Example value for epsilon_so
k_o = 1.89e-14      # Example value for k_o
DeltaP = 50000      # Example value for Delta P
mu = 0.00089        # Example value for mu
Rm = 1.37e11        # Example value for Rm
s = 0.00398

A = 1 - n_val - beta
Pa = (tau * np.exp((DeltaP / 1000) * theta)) * 1000
C = (2 * mu * s * A) / (k_o * Pa * (epsilon_so - s))

# Define the time interval and step
t_start = 100          # start time
t_end = 90 * 60        # end time in seconds (90 minutes)
dt = 100               # time step in seconds

t_values = np.arange(t_start, t_end + dt, dt)
y = np.zeros((len(t_values), 2))

# Loop through time steps
for i, t in enumerate(t_values):
    # Initial guesses for [DeltaPc, J]
    x0 = [100, 5e-5]

    # Define system of equations
    def F(x):
        return [
            x[0] - DeltaP + (mu * Rm * x[1]),
            x[1] + (((1 + (x[0] / Pa))**A - 1) /
                  np.sqrt(C * (((1 + (x[0] / Pa))**A) - 1) * t))
        ]

    # Solve using fsolve
    sol = fsolve(F, x0)
    y[i, 0] = sol[0] # DeltaPc
    y[i, 1] = sol[1] # J

# Extract results
DeltaPc_values = y[:, 0]
J_values = y[:, 1]

# Plot J vs. t
plt.figure()
plt.plot(t_values / 60, J_values * 1000 * 3600) # LMH conversion
plt.xlabel('Time (minutes)')
plt.ylabel('Permeate flux (LMH)')
plt.title('Permeate flux vs. Time')
plt.grid(True)
plt.show()
```

iv) Script for predicting the total pressure drop increase in a constant flux run

```
# Import python packages and libraries
import numpy as np
import matplotlib.pyplot as plt
from scipy.optimize import fsolve

# Define the parameters
n_val = 1.35
beta = 0.15
Pa = 1100
k0 = 1.89e-14
eps0 = 0.3
s = 0.00398
mu = 0.00089
J = 9.722e-6
Rm = 1.12e11

A = 1 - n_val - beta
C = (2 * mu * s * A) / (k0 * Pa * (eps0 - s))

# Define the time interval and step
t_start = 300           # seconds
t_end = 90 * 60         # 90 minutes in seconds
dt = 300                # step size in seconds
t_values = np.arange(t_start, t_end + dt, dt)

# Initialize cumulative P value
P_total = 0
y = np.zeros((len(t_values), 2)) # column 0 = P, column 1 = cumulative
P_total

for i, t in enumerate(t_values):
    # Define the function for fsolve
    def F(P):
        return P - ((((J**2) * C * t) + 1)**(1 / A) - 1) * Pa - (mu * Rm * J)

    # Initial guess
    P0 = 100

    # Solve for P
    P_solution = fsolve(F, P0)[0]
    # Accumulate
    P_total += P_solution

    # Store results
    y[i, 0] = P_solution
    y[i, 1] = P_total

# Extract cumulative total pressures
TotalP_values = y[:, 1]

# Plot results
plt.figure()
plt.plot(t_values / 60, TotalP_values / 1000) # Convert time to minutes,
pressure to kPa
```

```
plt.xlabel("Time (minutes)")
plt.ylabel("Total pressure drop (kPa)")
plt.title("Total pressure drop vs. Time")
plt.grid(True)
plt.show()
```

REFERENCES

1. W. Naessens, T. Maere, I. Nopens, Critical review of membrane bioreactor models – Part 1: Biokinetic and filtration models, *Bioresour. Technol.* 122 (2012) 95–106. https://doi.org/10.1016/j.biortech.2012.05.070.
2. H. Hamedi, O. Mohammadzadeh, S. Rasouli, S. Zendehboudi, A critical review of biomass kinetics and membrane filtration models for membrane bioreactor systems, *J. Environ. Chem. Eng.* 9 (2021) 106406. https://doi.org/10.1016/j.jece.2021.106406.
3. X. Cai, Z. Wang, Y. Qian, A. Wang, Y. Yang, S. Xia, Comprehensively understanding fouling properties of cake and bulk sludge in an anammox membrane bioreactor: Focusing on the composition, interfacial thermodynamics and microbial community, *J. Environ. Chem. Eng.* 10 (2022) 108612. https://doi.org/10.1016/j.jece.2022.108612.
4. B.K. Hwang, C.H. Lee, I.S. Chang, A. Drews, R. Field, Membrane bioreactor: TMP rise and characterization of bio-cake structure using CLSM-image analysis, *J. Memb. Sci.* 419–420 (2012) 33–41. https://doi.org/10.1016/j.memsci.2012.06.031.
5. L. Zhou, S. Xia, L. Alvarez-Cohen, Structure and distribution of inorganic components in the cake layer of a membrane bioreactor treating municipal wastewater, *Bioresour. Technol.* 196 (2015) 586–591. https://doi.org/10.1016/j.biortech.2015.08.026.
6. E. Poorasgari, T.V. Bugge, M.L. Christensen, M.K. Jørgensen, Compressibility of fouling layers in membrane bioreactors, *J. Memb. Sci.* 475 (2015) 65–70. https://doi.org/10.1016/j.memsci.2014.09.056.
7. M.K. Jørgensen, T.V. Bugge, M.L. Christensen, K. Keiding, Modeling approach to determine cake buildup and compression in a high-shear membrane bioreactor, *J. Memb. Sci.* 409–410 (2012) 335–345. https://doi.org/10.1016/j.memsci.2012.04.002.
8. R. Bai, C. Tien, Further work on cake filtration analysis, *Chem. Eng. Sci.* 60 (2005) 301–313. https://doi.org/10.1016/j.ces.2004.08.008.
9. J. Olivier, J. Vaxelaire, E. Vorobiev, Modelling of cake filtration: An overview, *Sep. Sci. Technol.* 42 (2007) 1667–1700. https://doi.org/10.1080/01496390701242186.
10. E. Vorobiev, Nonlinear consolidation model for the prediction of constant pressure expression of semi-solid materials, *Chem. Eng. Sci.* 235 (2021) 116458. https://doi.org/10.1016/j.ces.2021.116458.
11. K. Stamatakis, C. Tien, Cake formation and growth in cake filtration, *Chem. Eng. Sci.* 46 (1991) 1917–1933. https://doi.org/10.1016/0009-2509(91)80153-P.
12. C. Tien, R. Bai, B.V. Ramarao, Analysis of cake growth in cake filtration: Effect of fine particle retention, *AIChE J.* 43 (1997) 33–44.
13. Z. Zhu, Z. Wang, H. Wang, Y. Kong, K. Gao, Y. Li, Cake properties as a function of time and location in microfiltration of activated sludge suspension from membrane bioreactors (MBRs), *Chem. Eng. J.* 302 (2016) 97–110. https://doi.org/10.1016/j.cej.2016.05.031.
14. F.M. Tiller, C.S. Yeh, W.F. Leu, Compressibility of particulate structures in relation to thickening, filtration, and expression—A review, *Sep. Sci. Technol.* 22 (1987) 1037–1063. https://doi.org/10.1080/01496398708068998.
15. V. Pillay, *Modelling of Turbulent Cross-flow Microfiltration of Particulate Suspensions*, Department of Chemical Engineering, University of Natal, 1991.
16. J.F. Richardson, J.H. Harker, J.R. Backhurst, Liquid filtration, in: *Particle Technology and Separation Processes*, 5th ed., Butterworth Heinemann, Oxford, 2002: pp. 372–436. https://doi.org/10.1016/B978-0-08-049064-9.50018-7.
17. A.C.B. Neiva, L. Goldstein, A procedure for calculating pressure drop during the build-up of dust filter cakes, *Chem. Eng. Process. Process Intensif.* 42 (2003) 495–501. https://doi.org/10.1016/S0255-2701(02)00066-1.
18. F.M. Tiller, The role of porosity in filtration – numerical methods for constant rate and constant pressure filtration based on Kozeny law, *Chem. Eng. Prog.* 49 (1953) 467–479.
19. C. Tien, R. Bai, An assessment of the conventional cake filtration theory, *Chem. Eng. Sci.* 58 (2003) 1323–1336. https://doi.org/10.1016/S0009-2509(02)00655-3.

20. C. Tien, S.K. Teoh, R.B.H. Tan, Cake filtration analysis – The effect of the relationship between the pore liquid pressure and the cake compressive stress, *Chem. Eng. Sci.* 56 (2001) 5361–5369. https://doi.org/10.1016/S0009-2509(01)00263-9.
21. F.M. Tiller, W.F. Leu, Basic data fitting in filtration, *J. Chinese Inst. Chem. Eng.* 11 (1980) 61–70.
22. D.J. Lee, C.H. Wang, Theories of cake filtration and consolidation and implications to sludge dewatering, *Water Res.* 34 (2000) 1–20. https://doi.org/10.1016/S0043-1354(99)00096-2.
23. J. Wu, C. He, X. Jiang, M. Zhang, Modeling of the submerged membrane bioreactor fouling by the combined pore constriction, pore blockage and cake formation mechanisms, *Desalination* 279 (2011) 127–134. https://doi.org/10.1016/j.desal.2011.05.069.
24. A.E. Farooqui, A.M. Pica, P. Marocco, D. Ferrero, A. Lanzini, S. Fiorilli, J. Llorca, M. Santarelli, Assessment of kinetic model for ceria oxidation for chemical-looping CO_2 dissociation, *Chem. Eng. J.* 346 (2018) 171–181. https://doi.org/10.1016/j.cej.2018.04.041.
25. M.F. Mohamad Yusop, A.Z. Abdullah, M.A. Ahmad, Amoxicillin adsorption from aqueous solution by Cu(II) modified lemon peel based activated carbon: Mass transfer simulation, surface area prediction and F-test on isotherm and kinetic models, *Powder Technol.* 438 (2024) 119589. https://doi.org/10.1016/j.powtec.2024.119589.
26. O.O. Ogunbiyi, N.J. Miles, N. Hilal, The effects of performance and cleaning cycles of new tubular ceramic microfiltration membrane fouled with a model yeast suspension, *Desalination* 220 (2008) 273–289. https://doi.org/10.1016/j.desal.2007.01.034.
27. A. Valencia, C. Le Men, C. Ellero, C. Lafforgue-Baldas, J.F. Morris, P. Schmitz, Direct observation of the microfiltration of yeast cells at the micro-scale: Characterization of cake properties, *Sep. Purif. Technol.* 298 (2022) 121614. https://doi.org/10.1016/j.seppur.2022.121614.
28. E. Akhondi, F. Wicaksana, A.G. Fane, Evaluation of fouling deposition, fouling reversibility and energy consumption of submerged hollow fiber membrane systems with periodic backwash, *J. Memb. Sci.* 452 (2014) 319–331. https://doi.org/10.1016/j.memsci.2013.10.031.
29. V. Jegatheesan, B.K. Pramanik, J. Chen, D. Navaratna, C.Y. Chang, L. Shu, Treatment of textile wastewater with membrane bioreactor: A critical review, *Bioresour. Technol.* 204 (2016) 202–212. https://doi.org/10.1016/j.biortech.2016.01.006.
30. L. Böhm, A. Drews, H. Prieske, P.R. Bérubé, M. Kraume, The importance of fluid dynamics for MBR fouling mitigation, *Bioresour. Technol.* 122 (2012) 50–61. https://doi.org/10.1016/j.biortech.2012.05.069.
31. S. Chang, Application of submerged hollow fiber membrane in membrane bioreactors: Filtration principles, operation, and membrane fouling, *Desalination* 283 (2011) 31–39. https://doi.org/10.1016/j.desal.2011.03.025.
32. U.T. Rahman, H. Roy, M.R. Islam, M. Tahmid, A. Fariha, A. Mazumder, N. Tasnim, M.N. Pervez, Y. Cai, V. Naddeo, M.S. Islam, The advancement in membrane bioreactor (MBR) technology towards sustainable industrial wastewater management, *Membranes (Basel)* 13 (2023) 1–28.
33. D.C. Sioutopoulos, A.J. Karabelas, The effect of permeation flux on the specific resistance of polysaccharide fouling layers developing during dead-end ultrafiltration, *J. Memb. Sci.* 473 (2015) 292–301. https://doi.org/10.1016/j.memsci.2014.09.030.
34. G. Singh, L. Song, Cake compressibility of silica colloids in membrane filtration processes, *Ind. Eng. Chem. Res.* 45 (2006) 7633–7638. https://doi.org/10.1021/ie060481g.
35. Y. Gao, X. Shi, X. Jin, X.C. Wang, P. Jin, A critical review of wastewater quality variation and in-sewer processes during conveyance in sewer systems, *Water Res.* 228 (2023) 119398. https://doi.org/10.1016/j.watres.2022.119398.
36. Y. Duan, H. Xi, Z. Qin, R. Guo, F. Wang, Y. Yuan, Water quality characteristics of municipal wastewater treatment plants and the prospect of reclaimed water utilization in lower-middle income and water-scarce areas: A case study of Puyang, *Water Cycle* 6 (2025) 61–70. https://doi.org/10.1016/j.watcyc.2024.11.002.
37. T. Gehring, E. Deineko, I. Hobus, G. Kolisch, M. Lübken, M. Wichern, Effect of sewage sampling frequency on determination of design parameters for municipal wastewater treatment plants, *Water Sci. Technol.* 84 (2021) 284–292. https://doi.org/10.2166/wst.2020.588.

Index

For Product Safety Concerns and Information please contact our
EU representative GPSR@taylorandfrancis.com Taylor & Francis
Verlag GmbH, Kaufingerstraße 24, 80331 München, Germany